# Defying Gravity

W. W. NORTON & COMPANY NEW YORK LONDON

# Defying Gravity

## LAND DIVERS, ROLLER COASTERS, GRAVITY BUMS, AND THE HUMAN OBSESSION WITH FALLING

Garrett Soden

Frontispiece: Photograph by Patrick Krohn,
www.krohn-photos.com.


Printed in the United States of America
First published as a Norton paperback 2005



Manufacturing by The Courier Companies, Inc.
Book design by Chris Welch
Production manager: Amanda Morrison

LIBRARY OF CONGRESS CATALOGING-IN-PUBLICATION DATA
Soden, Garrett.
[Falling]
Defying gravity : land divers, roller coasters, gravity bums, and the human obsession with falling / by Garrett Soden.
p. cm.
Includes bibliographical references and index.
ISBN 0-393-05413-6 (hardcover)
1. Daredevils—History. 2. Jumping—Psychological aspects. 3. Extreme sports. I. Title.
GV1839.S633 2005
797.5—dc22

2004025111

**ISBN 0-393-32656-X (pbk.)**

W. W. Norton & Company, Inc., 500 Fifth Avenue, New York, N.Y. 10110
www.wwnorton.com

W. W. Norton & Company Ltd., Castle House, 75/76 Wells Street, London W1T 3QT

1 2 3 4 5 6 7 8 9 0

This book is dedicated, with all the love I am capable of, to Kate Shein.

# Contents

# Defying Gravity

The improbable evolution of bungee jumping from a South Pacific native ritual, shown here, to an amusement ride for tourists is just one of the extraordinary ways in which the sensation of falling has been popularized over the last few decades, continuing a process that actually started more than two centuries ago. *(Photograph by Patrick Krohn, www.krohn-photos.com)*

INTRODUCTION

# Birth of the Bungee Jump

People may think we are mad. We think they are insane to endure such humdrum lives. —DAVID KIRKE, WORLD'S FIRST BUNGEE JUMPER

NATIVES WHO JUMP from hundred-foot towers headfirst into dirt. Sure. Tell us another.

That was Irving and Electra Johnson's reaction to Oscar Newman's tale about what he called "land divers." The year was 1952, and the Johnsons, who for a decade had sailed the world reporting for *National Geographic*, were always on the lookout for wild stories, as their friend well knew. But this was too much.

Yet Newman swore it was true. On the tropical island of Pentecost, one of the Vanuatu Islands located thirteen hundred miles east of Australia in the South Pacific, land diving was an ancient ritual. Newman had seen it himself. He had lived on Pentecost for years.

What saved the land divers from death, he said, was that the dirt was tilled until it was very soft, and more importantly, the divers attached long vines to the tower and tied them around their ankles to break their fall. "Let me know when you're coming back,"[1] he told the Johnsons; he would arrange a demonstration.

When the Johnsons arrived during their 1953–55 voyage, the islanders had been expecting them; the men even wore shorts instead of their usual dangling patch of matted leaves, so as not to offend Western modesty. They had also spent days preparing the tower. First, they had selected a tree that looked sturdy enough to form the backbone of their tower. Next, they had stripped off most of its branches and cut down the surrounding jungle to make a clearing. To the main tree they had lashed hundreds of branches and trunks, forming a leaning, springy cone of lattice work eighty feet high. Tied to the back of the cone were vines stretched like guy wires to the stumps of the trees they had felled. At the tip of the cone, a platform that looked like a short diving board jutted out; below it, other platforms stuck out at various heights.

All the jumpers would be male; the ritual, in fact, was designed to both impress and defy women. Some time long ago, the natives told the Johnsons, a man named Tamalié became furious when his wife took a lover. He chased her to the top of a palm tree, where she tied vines to her ankles. When Tamalié lunged at his wife, she jumped and he fell. She was saved; he died. Since then, according to legend, the men thought it would be a good idea if they practiced the stunt just in case they found themselves in the same predicament. As a further precaution, they banned women from the ritual. Females weren't allowed anywhere near the tower as it was being erected, and certainly weren't allowed to jump when it was finished. What they *were* required to do during the jumps was to stand silently and listen as the men made speeches from the platform, most of which detailed their wives' bad behavior.

You didn't have to be married, though, to jump. Each year the ceremony began with the least-experienced jumpers and progressed until the veterans tried for the most awesome leaps. The first to go when the Johnsons were there was a boy of eight, who sailed out from a platform twenty-five feet above the dirt mound.

Before each jump, women dancers in grass skirts and bare to the waist chanted and swirled, becoming more excited as the men climbed higher. Each jumper had his own style, milking his time on the platform with songs, clapping, slow-motion pantomimes, dramatic posturing, and speeches. Some pretended to lose their balance and nearly fall. Sometimes their courage would fail, and they would simply climb back down. The jumping continued for six hours.

The last jumper that day was Warisul, handsome and slim, who stood on a long board at the top of the cone nearly eighty feet above the ground. On a jump like this the leaper would be falling at about forty-five miles per hour by the time the vines and the dirt stopped him, and so preparation was critical. The dirt mound was freshly sifted and fluffed. The vines were measured to take into account their stretch, which wasn't much, as well as the flex of the tower and the braking action of the diving board itself, which snapped its supporting props when the weight of the jumper jerked the vines taut. Crude as the system seemed, it worked. No one was hurt that day, and many of the jumpers actually sprang back onto their feet from the recoil of the vines and tower.

But there could also be close calls because of all the variations in the vines and tower and because it mattered how a jumper jumped. If he didn't dive out far enough, he would hit the sloping ground too high and break his neck. If he dove too far, he would snap the vines, with the same result. Warisul's jump was just about perfect. He dove out in a clean arc, well away from the tower, plummeting headfirst, with the vines trailing in slack coils, then straightening and stretching. When he was just a few feet above the ground, the vines jerked the platform down, crushing its props, then bent the whole shuddering tower like an enormous spring, just as Warisul's head struck the ground. One vine snapped, but he was yanked back into the air and came down on his feet, where he stood smiling as the women rushed in to splash him with cool water.

IF THE ISLANDERS' ritual hadn't been a gravity stunt—if it had been, say, an unusual dance, like the hula—the Johnsons probably wouldn't have gone out of their way to see it, nor would they have devoted more than a few paragraphs to it, as they had to other curiosities in previous articles such as "*Yankee* Roams the Orient" and "Westward Bound in the *Yankee*."

But the sight of people flinging themselves from a high place mesmerized its spectators. So the article that appeared in 1955 was titled "South Seas' Incredible Land Divers," and all sixteen pages of it—including twenty-two spectacular photographs (culled from more than sixteen hundred taken that day)—focused entirely on this newly discovered ritual. As word spread, a small but growing stream of tourists went out of *their* way to see the islanders' remarkable leaps.

One of these new visitors, arriving a dozen years after the Johnsons, was Kal Muller, a journalist in his twenties. Muller would also report on the land divers for *National Geographic.* But it was now the late 1960s, and times had changed. Along with sex, drugs, and rock 'n' roll, the thrill of gravity—mostly in the form of surfing and rock climbing at Yosemite—was fueling a new kind of rush for the most adventurous among Muller's generation. Muller, in fact, had done some skydiving. Unlike the Johnsons, he wouldn't be content to simply gawk. He wanted to dive.

And he wanted to experience land diving in its purest form. Rather than visit the most accessible Pentecost villagers, who by now were performing for tourists, Muller chose Bunlap, the extremely remote tip of the island where the natives still wore the traditional grass skirts and dove only for their own ritual purposes. Wary of outsiders (they were so isolated that when Muller showed a film, a youngster asked "if little people were performing behind the viewing screen"), the villagers stalled for two years before they agreed to let Muller participate.

Muller lived with the people of Bunlap for seven months before the jump, gaining their trust and well aware of his honorable position as the first white man ever to be allowed to perform the feat. On the day of his dive, he climbed to a platform fifty feet above the ground. He was ready to go when the village chief reminded him that he should make a speech. Muller spoke in the local pidgin English. "Me fella," he began, "me glad too much belong me stop with em you allgetta. Me learn em plenty someting longcustom belong you fella. Me like em you fella too much.

Now here me fella glad too much you allgetta you let em me jump long land dive." He then clapped his hands three times over his head and pushed off.

Sailing down, he kept his knees bent and his arms wrapped to his chest to avoid breaking them on impact. He had been told to keep his eyes closed, but he opened them to see the ground rushing up. His head had just barely touched the dirt when the vines yanked him back, leaving him hanging upside down. The villagers shouted their approval, rushed in, and cut him down. "Me look you no fright," one told him.[2]

## The Dangerous Sports Club

Published in 1970, Muller's account of his own land diving would provide the detonator for the next blast in the chain reaction that would lead to the invention of bungee jumping. This one would be set off nine years later by the kind of brazen characters who regularly inflame public interest in gravity feats. And even by the wild standards of that history, the crew that would actually invent the new sport would seem crazier than most.

They met on a Swiss mountainside in 1977, when Chris Baker was approached by David Kirke and Ed Hulton, who asked to borrow Baker's hang glider for a short flight. Afterward, the three retired to a local bar to raise a few warm pints. They had much in common, it turned out. Like Baker, Kirke and Hulton were upper-crust Brits. The two had met at Oxford in the sixties, and since then they hadn't done much more than try to find ways to postpone proper adulthood. Baker understood.

All of them had a craving for excitement and hadn't found much of it in traditional sport, which is why they had been attracted to hang gliding. Something must be done, they agreed. Existing sports had to be pushed to new limits, and—even better—new sports that delivered a bigger kick had to be invented. They would be a club—the Dangerous Sports Club, one of them dubbed it, a name that was satirical on the face of it because they were hardly organized enough to be a club, and much of what they would do would have nothing to do with danger or sports, unless you consider drinking a risky athletic event.

Over the next two years, the club consisted of whoever showed up at one of their bizarre events, one of which nearly did make drinking a dangerous sport. This was a cocktail party they held on Rockall, a shark fin–shaped stone slab that jutted seventy feet out of the Atlantic two hundred miles off the north Scottish coast. They set out on a summer's day from Crinan, in a small boat overloaded with friends, food, and alcohol. As they left the smooth water of the Crinan Canal, the wind kicked up to a force-nine gale.

After two days out in the storm, the heaving boat began to leak. Clinging to the rail, waist deep in stinking, freezing seawater, the passengers began to vomit on each other. They plugged the leak with a cork from a champagne bottle. After three more days, they reached Rockall, scrambled up the slanted stone face and ate and drank for the day. When they were done, Kirke and Baker leapt from the seventy-foot cliff into the ocean, scrambled into their boat, and sailed home.

In later years, the club would hold such eccentric events as a ski race in which members slid down a run at Saint Moritz in vehicles that included a bathtub, an ironing board, a rowboat, a grand piano (played by a drunken member as it slid), a Louis XIV dining suite, and a toilet.

By 1979, although they had also done some whitewater kayaking and had hang glided from the summit of Mt. Kilimanjaro, the club hadn't invented any new sports—yet.

THE MOST FAMILIAR local landmark for Chris Baker, who lived along the lush banks of the Avon River in Bristol, England, was the massive Clifton Suspension Bridge that spanned the 250-foot-deep gorge, its Egyptian-style towers rising from cliffs thick with trees. One day, as Baker was packing up for a hang-gliding trip, he looked up at the bridge and suddenly remembered a plan Hulton had proposed earlier. Hulton, it turned out, had seen the articles in *National Geographic.* He had suggested at the time that land diving was the perfect sport for the club—but no one was prepared to go all the way to Pentecost, so the idea was dropped.

Baker again looked up at the bridge, then at what he was holding: the cords he was using to lash his glider to the car. And it came to him. Bungee cords. Instead of vines, use bungee cords—*enormous,* long bungee cords. And instead of diving from a tower on Pentecost, dive from the Clifton Suspension Bridge right here in Bristol.

Hulton and Kirke thought it was madness—a *grand* madness. Baker had found cords he thought would work. Manufactured to catapult gliding planes into the air, they surely would support a man's weight. At least that was the theory. And a theory is all they would have. They agreed not to try any tests—that wouldn't be sporting.

They would go on April Fools' Day. On the night of March 31, 1979, Baker threw a raucous all-night party in anticipation of the morning's jump. By dawn, fueled by gallons of alcohol and plenty of hallucinogenic mushrooms, the crowd had partied themselves into a stupor.

Somehow the police had been tipped off that an illegal stunt was planned, and so the club hunkered down, waiting for the cops on the bridge to grow bored and leave. In the meantime, Baker left to fetch his girlfriend from London, some hundred miles away. While he was gone, the cops concluded that their tip was an April Fools' hoax, and left. Someone gave the all-clear sign. In spite of being hungover or still stoned—or perhaps because of it—a crew of four jumpers, dressed in top hats and tails, wiggled into climbing harnesses, gathered up their bungee cords, and piled into cars.

They emerged from a nearby garage as a caravan and headed to the bridge. When they arrived, Kirke climbed out of a friend's Renault and made for the rail. Someone tied the cords to the bridge. Kirke stepped over the balustrade, clutching a bottle of champagne in his left hand. Without hesitation, he tipped over, back first, and the crowd burst into howls and cheers. Down he sailed, arms akimbo. The others peered anxiously over the rail. He fell like a stone, and in his formal attire looked like an undertaker perhaps arranging for his own funeral. More than a hundred feet below them now, he practically vanished against the dark river. Then, halfway to his doom, the cord ran out of slack and began to stretch. He slowed—and then began to rise. It was working.

The champagne bottle was ripped from his hand. His top hat plummeted to the river. Kirke shot up toward the onlookers, flying to within twenty feet of them, then down again for the stretch, then up, bobbing like a wild marionette. Alan Weston and Simon Keeling went next, dropping almost simultaneously. A whistle shrieked—the police. With the cops racing to stop this lunacy, Tim Hunt let himself down over the rail, hung by his fingers, and let go. Below, a driver skidded his car off the road at the sight, then jumped out and stared, slack-jawed, upward.

All four were arrested for disturbing the peace and fined a hundred pounds each—a small price for becoming instantly famous. Fame was ensured because Kirke—an exhibitionist to his core who had emerged as the club's de facto leader—had notified the *Daily Mail* so the paper would have a photographer on the spot, and had also made arrangements with an independent filmmaker.

"David," the filmmaker had asked before the jump, "can you tell me what the Dangerous Sports Club is, exactly?"

Kirke could.

"It's a collection of people who are quite unacceptable in any other terms."

Their aim in life?

"To avoid everybody who has an aim."

Was this for some sort of financial reward?

"We wouldn't know what finance is, much less a reward."[3]

As far as the press was concerned, the Dangerous Sports Club was gold. The *Daily Mail* ran on its cover a half-page photo of the four plunging madmen in top hats and tails. A few days later, Kirke appeared on the BBC television show *Nationwide*.

Kirke's knack for publicity—a key ingredient that has always driven gravity play—ensured that the club's coverage wouldn't end with this maiden jump. "We wanted to start a craze,"[4] Kirke told one reporter. Kirke proved his instinct for promotion with his next move. Rather than repeat the jump in Britain, the club set its sights on the international capital of crazes—California—and at a venue where their performance would be impossible to ignore: the Golden Gate Bridge.

Six months after the Clifton jump, the crew made its assault on the pride of San Francisco, again dodging the authorities. The jumpers were Chris Baker, Alan Weston, Simon Keeling, Peter Carew, and for the first time, a woman: Janie Wilmot. They all sailed off the bridge—all except Baker. Ironically, after inventing bungee jumping and then missing its premier, Baker botched his second chance when his cords became tangled in the bridge's girders (he would finally make his first jump at a show in Cheltenham, England, in 1985). Keeling and Weston made their getaway aboard a boat bobbing in the chop of the bay; the others were arrested for trespassing and fined ten dollars each.

The American press gave the cheeky stunters enormous coverage. Within weeks, television producer Alan Landsberg came knocking on their door offering twenty thousand dollars if they would appear on his show *That's Incredible,* broadcast weekly on ABC with upward of twenty million viewers. And Landsberg had a great idea as to where to stage their antics: the highest bridge in the world, the Royal Gorge Bridge in Colorado, which spanned a chasm more than a thousand feet deep.

The club and its entourage happily obliged. They crammed themselves into borrowed and rented cars and set off from San Francisco to the Rocky Mountain state. This time Kirke jumped along with the others, and again acting as the club's pioneer, used cords longer than any of them had used before: a whopping 420 feet. The jump went fine; the recovery of the jumpers didn't. Hauling them back up to the bridge was far harder with their longer cords, and Kirke was left hanging in the cold November wind for hours. By the time he was pulled up, the tight climbing harness around his thighs had cut off circulation to his legs and he was in a state of near hypothermia. It was the first warning that bungee jumping, even when the equipment performed perfectly, might actually have some unforeseen danger.

Nonetheless, they had made a good chunk of money (all of which they spent on party supplies before ever leaving the state). And after only three jumps—thanks to Kirke's leveraging of publicity at each step—the club, and bungee jumping, were now known to millions. (The clip of the Royal Gorge jump was aired repeatedly.) As surprised as anyone, it looked like the motley gang of eccentrics had, as it were, fallen ass-backward into "financial reward" after all.

But with the initial shock value of the bungee fully exploited, the media moved on. The club would occasionally get coverage with other escapades, especially with their whimsical ski races (and the parties afterward, during one of which a club member swung from a Swiss hotel's chandelier, yanking it from the ceiling and sending it crashing down on top of him—all of which just happened to be caught on film). But their bread and butter soon became bungee jumping at store openings, carnivals, and the like, usually jumping from cranes hired for the occasion.

Even so, bungee jumping was still years away from becoming public recreation, and for good reason: it wasn't long before people started getting hurt. The first to be injured was Martin Lyster, who was smacked in the eye by a knotted rope attached to his harness and left with slight but permanent damage to his vision. In 1983, Kirke broke his leg jumping. That same year, Hugo Spowers, in trying to graze the ground for dramatic effect, used far too much cord and broke both legs and his pelvis. With its usual perverse humor, the club adopted a wheelchair as its logo.

As much as they made a show of being lunatics, and in spite of some accidents (usually the result of a jumper's bad decision, such as Spowers's), those in the club responsible for setting up the equipment were actually careful as well as knowledgeable; many were physics students educated at Oxford. Their imitators were often far less competent.

In 1986, the BBC hired an escape artist with no bungee experience to set up a jump for Noel Edmonds's *Late Late Breakfast Show*, which subjected audience volunteers to all manner of stunts. During a rehearsal, when a young bricklayer leapt from a crane, the TV crew watched in horror as they saw the bungee cords, unattached, trail through the air behind him. He was killed instantly when he hit the ground. An inquest was held (with club members testifying as experts for the prosecution), the BBC fined, and the show permanently canceled. After all the bad press, bungee jumping was dismissed by the public and authorities alike as no more than a bizarre and stupidly dangerous sideshow, like being shot from a canon. Everyone, it seemed, was relieved to see the shows die out.



## Building the Bungee Empire of A. J. Hackett

But, of course, this was not the end of bungee jumping. Although the Dangerous Sports Club had managed to invent only one new sport, it was a sport that delivered an unparalleled visceral kick—a kick that would prove to be a siren call for a significant portion of the population. These people wouldn't be deterred—in fact, they *couldn't* be deterred, because for them the quest for visceral sensation isn't a choice. It is a need—almost an instinctual drive, like hunger and sex—built into their very neurology, confirmed only a few years ago by the discovery of what has been called the "thrill gene." And falling, or the sustained, moment-to-moment risk of falling (as in rock climbing), is probably the most efficient way to satisfy that need. The man who intuitively understood this and who would become determined to make this peculiar form of falling safe enough for mass consumption was, in 1985, a handsome twenty-seven-year-old New Zealander named A. J. Hackett.

As a member of his country's speed skiing team, Hackett was already well acquainted with the rush gravity could generate when, several years earlier, he had seen an amazing new way to get it: the film of the Dangerous Sports Club's leap from the Royal Gorge Bridge. But unlike other early bungee jumpers, he didn't run out and jump off a bridge with bungees borrowed from the military. Hackett owned a construction company; he knew that anything meant to support human weight had better be failure-proof. So he decided to design his own bungee cords from scratch.

Hackett and his friend Chris Sigglekow somehow convinced a professor at Auckland University to help them develop a formula for rubber cords that would yield predictable results. Toward the end of 1986, after rigorous testing in the lab, they ventured out to Greenhithe Bridge, where—if something went wrong—they would plunge into deep water from a drop of only sixty feet. And even then, they tested the cords with sandbags before jumping themselves. All that care resulted in exactly what they wanted: a thrilling, safe jump.

Like other bungee entrepreneurs of the time (notably John and Peter Kockelman in America), Hackett began by offering jumps set up just for the occasion, usually from bridges. These jumps, though, required customers to drive and often hike to the jump site; an outing could easily consume the entire day. Along with new partner Henry van Asch (another speed skier—van Asch held the New Zealand record, at over one hundred miles per hour), Hackett saw that this model—essentially a customized tour operation—would never grow beyond a small business. Hackett and van Asch then hit upon a completely different vision: they would reinvent bungee jumping as an amusement ride.

Safety would be paramount, and on that score, they were already way ahead. They had bungee cords that performed far more predictably than others, so predictably that in 1987 Hackett had jumped from a cable gondola three hundred feet above a mountain slope and just gently touched the snow bank below before rebounding—exactly as he had planned. They had also learned how to handle a crowd of jumpers, as a result of bungee parties they had held.

The pair opened the world's first permanent bungee site on the Kawarau Bridge near Queenstown, New Zealand, in November of 1988. It was an immediate success. They went on to open a new site almost yearly, thanks in no small part to van Asch's experience as a professional sports promoter: in 1989, they opened a second site at Queenstown; in 1990, sites in Normandy, France, and Cairns, Australia; in 1994, Las Vegas, Nevada; in 1995, Kuta Beach, Bali; in 1997, a third site in Queenstown; in 1998, Nuremberg, Germany; in 1999, Acapulco, Mexico, and a fourth site in Queenstown. By 1999, AJ Hackett Bungy (their preferred spelling) had dropped over one million jumpers.

And incredibly, they came up with ways to make bungee jumping *more* thrilling. Today, in addition to the plain vanilla jump, which costs about sixty dollars, you can launch into a bungee drop from a skateboard, mountain bike, diving board, water slide, or helicopter.

Along the way, the company almost single-handedly transformed Queenstown from a ski and boating resort into a virtual gravity playground. Companies riding Hackett's coattails now offer wire sliding and cable swinging over canyons, as well as skydiving, hang gliding, parasailing, and abseiling.

As of this writing, AJ Hackett Bungy plans to open a three-million-dollar bungee complex at the original Kawarau Bridge site in the summer of 2002. It is being built largely underground so as not to disturb the canyon's beauty and will house a viewing deck, cafe, bar, retail shopping (featuring AJ Gear clothing), interactive displays, a bungee museum, and "a 15-metre high domed screen theatre, offering visitors a 'Cinematic Bungy' experience,"[5] as the company's press release puts it.

Meanwhile, back on Pentecost Island, land-diving performances became so swamped with tourists that the government had to shut them down for an entire year in 1995.

## Gravity Battles

The success of a weird sport like bungee jumping could be considered an aberration if it weren't for the fact that we now seem surrounded by dozens of such weird sports, as well as radical thrill rides based on the same appeal. This mania to find different ways to stimulate our sense of gravity is relatively new, beginning about twenty years ago.

Vertical (or "vert") skateboarding was born when kids began flying above the lips of empty swimming pools around 1979; BMX bike riders went vert about the same time. By the mid-1980s, snowboarders were turning flips above snow half-pipes, while radical surfers "got air" by rocketing off the lips of waves. Sports already based on falling or risking a fall became more intense. Rock climbing routes up sheer cliffs that had been rated impossible (a "10") were conquered—so many increasingly "impossible" routes were completed, in fact, that the scale had to be adjusted (today, the most difficult climb stands at 14). Skydivers began turning flips while strapped to the newly invented skysurfing board; they also invented BASE jumping—the practice of parachuting off high structures or cliffs. ("BASE" is an acronym for the objects of the jumpers' lust: buildings, antennaes, spans, and earth.) Roller coasters not only began to loop, corkscrew, and steepen (to eighty-five degrees—nearly straight down) but also set records throughout the late 1980s for dropping riders farther: 138 feet in 1987; 155 feet in 1988; 194 feet in 1989; and 228 feet, nearly the drop from the Golden Gate Bridge, in 1991. And, of course, bungee jumping spread worldwide.

This mass appeal of gravity play, though, didn't appear without foreshadowing. Skateboarding, snowboarding, and skysurfing all grew from the surfing craze that began in the early 1960s. Today's thrill-ride boom was presaged by one in the 1920s, the Golden Age of Roller Coasters, when more than a thousand across America were designed specifically to deliver the "sensationally deep and thrilling dips"[6] demanded by the public, according to coaster designer John Miller. Extreme rock climbing was pioneered in Yosemite from the 1950s and can trace its aesthetics back to the Golden Age of Mountaineering, which peaked in the 1860s. Even the Dangerous Sports Club and its media-hyped jumps were preceded by the flamboyant, headline-grabbing daredevils of the 1800s who leapt from cliffs, jumped from the Brooklyn Bridge, or walked on tightropes over Niagara Falls.

The real shift toward our gravity-friendly attitude, in fact, didn't happen two decades ago but about two centuries ago. That's when the first man dropped from a parachute, when the roller coaster and circus were invented, and when people first started climbing mountains for recreation.

But the shift didn't occur without a fight. Just the idea of falling, after all, had been a powerful negative metaphor for millennia. Human culture is filled with stories that equate falling with failure: Icarus fell from the Greek sky because he lacked humility; Lucifer fell from Christian heaven to become Satan; we say that leaders fall from power, that civilizations fall into barbarism, that sinners fall from grace. (Even the phrase *falling in love* implies a loss of control, as does *to fall asleep* or *to fall under a spell*; *Miriam Webster's* dictionary defines this usage as "to pass suddenly and passively" into these states.) So over the last two centuries, every time a new gravity activity became popular, it would be attacked—usually on moral grounds that echoed the ancient metaphors and often in terms that bordered on hysteria. To pick just two examples, in the early eighteenth century a New Jersey newspaper warned townspeople who had seen high-diver Sam Patch that they had "incensed God" and "are daring the vengeance of an offended Deity,"[7] while in 1865 the London *Times* condemned mountaineering, asking, "Why is the best blood of England to waste itself in scaling hitherto inaccessible peaks, in staining the eternal snows and reaching the unfathomable abyss never to return? . . . Is it duty? Is it common sense? Is it allowable? Is it not wrong?"[8]

Throughout all the battles, one thing has remained clear: the sensation of falling stirs up intense emotions. Falling can feel like rapture, or if you're taken unwillingly, like rape. People violently disagree about avoiding its dangers and exploring its pleasures—and even about what those dangers and pleasures are. Free-soloist rock climbers—those who clamber up thousand-foot cliffs without ropes—will tell you that being on the knife-edge of the sensation makes them feel alive the way no other experience can. Their critics—including some other climbers—say free soloists are simply insane.

The strong emotions that falling generates has filled history with strong characters: daredevil parachutists who taught the army how to save pilots; Niagara Falls barrel riders who often killed themselves pursuing fame; a high diver who enjoyed a career built on a jump he never made; early stuntmen who would try "any foolish stunt my psychopaths in the writing room could think up,"[9] as director Mack Sennett put it; California street kids who broke into backyards to skate empty swimming pools and who invented modern skateboarding; and young climbing bums in Yosemite who drank, smoked dope, and slept in ragged tents by night but who by day were the best rock climbers the world had ever seen.

It is also a history filled with discoveries: that dragons didn't live in the Alps, and so the Alps could be climbed; that humans could free-fall for thousands of feet without becoming unconscious, as was thought; that there was an enormous audience for extreme sports, as ESPN discovered after broadcasting the first X Games; and most significantly, that there was a real difference in the neurology of people who liked falling and those who didn't, and that this difference revealed plenty about both types.

Ultimately, this is a story of how we come to terms with human desire. In this case, the desire to resist, control, and enjoy falling. It's a desire that can't be separated from our sense of gravity, a sense so neglected it isn't included in the traditional five senses—even though it was among the earliest to evolve and helped build virtually every part of our body, from our gripping hands to our thinking brain.

How we feared, vilified, and eventually celebrated the sensation of falling has a lot to say about what makes us human.

# PART I

# Daredevils, Heroes, and Madmen

The world's first parachute jump, made by André Jacques Garnerin in Paris in 1797, capped a century of expanding interest in gravity play that included the invention of the circus, the roller coaster, and mountaineering as a form of recreation. The illustration above shows Garnerin's first jump in England, in 1802.

CHAPTER 1

# The Gravity Century

I was on the point of cutting the cord that suspended me between heaven and earth . . . and measured with my eye the vast space that separated me from the rest of the human race. . . . I felt myself precipitated with a velocity which was checked by the sudden unfolding of my parachute. . . . At length I perceived thousands of people, some on horseback, others on foot, following me, all of whom encouraged me by their wishes, while they opened their arms to receive me.

— ANDRÉ-JACQUES GARNERIN, WORLD'S FIRST PARACHUTIST

IT WAS AN autumn afternoon in Paris, October 22, 1797—cloudless and perfect. In Monceau Park, on the outskirts of the city, a thousand people swarmed around André-Jacques Garnerin, who had made the astonishing announcement that he would be the first person in history to fall several hundred feet from a balloon using a contraption called a parachute. Most were skeptical; plenty were there, no doubt, simply to trade sarcastic remarks about this character.

Garnerin could have hardly expected better. The problem was that he had made the same astonishing announcement only five months earlier and had sold hundreds of tickets to a performance that had collapsed into a near riot when he couldn't even get his balloon off the ground, never mind the parachute. The furious crowd had berated him. The newspapers had had a field day at his expense.

In spite of this earlier failure, a sizable crowd had once again gathered after hearing about this latest spectacle; over the past several years, Parisians had become accustomed to looking skyward as all manner of astounding new things had begun appearing overhead.

The first had been the hot-air balloon. Only fourteen years before, papermakers Joseph and Étienne Montgolfier had begun experimenting with paper globes filled with hot air, which, when they drifted into the provinces, terrified locals to such an extent that authorities had to calm the nation with official reassurances that the balloons were harmless. The commotion had piqued the interest of the court, and that same year the Montgolfiers found themselves at Versailles demonstrating their invention for the pleasure of King Louis XVI and Marie-Antoinette—this time, though, with the balloon lifting a basket that held a sheep, a duck, and a rooster.

The next logical step had been to see if men could fly, and after at first considering sending up a pair of criminals—no great loss if they perished—it was decided that the honor instead should go to upstanding citizens. Jean-François Pilâtre de Rozier, a scientist, and the Marquis d'Arlandes eagerly volunteered, and only months after the animals had gone up, the two men stepped into the Montgolfier brothers' extraordinary new invention. Alternately stoking the fire in their balloon's basket to lift it and swatting flying embers with wet sponges to keep their craft from exploding in flames, the pair managed to reach an altitude of three thousand feet and traveled nearly five miles before they set down—a success that served notice to the public that they should watch the skies. Shortly after, when Jacques Charles went up in the first hydrogen balloon, half of Paris—400,000 people—turned out to watch.

In the meantime, in Montpellier, it so happened that Louis-Sébastien Lenormand, professor of technology at the Paris Conservatory of Arts and Handicrafts, had just successfully tested his own new invention with a basketful of animals: he had dropped them from a tower protected by something he called a *parachute*—a word of his own coinage. Not to be upstaged, Joseph Montgolfier stole Lenormand's design and started dropping animals from towers as well. Then, stealing both ideas, showman Jean-Pierre Blanchard combined the balloon and the parachute for profit and toured the country with his act of taking livestock skyward in his balloon and then sending them floating back to earth beneath a canopy.

So the parachute, while still a novelty, wasn't new. Up to this point, though, no man had tried risking one of these drops himself, and considering his demonstrated incompetence as a balloonist, few believed that Garnerin had the qualifications to be the first.

It couldn't have been a good sign that Garnerin's assistant looked petrified. ("I made all my efforts to dissuade him to try this perilous enterprise,"[1] he would write later.) As he and Garnerin wrestled with the limp balloon that was inflating at an infuriatingly slow rate, spectators got a chance to examine the apparatus in which Garnerin would shortly risk his life. Lying on the ground next to the balloon, it looked exactly like a gigantic folded umbrella. Two-dozen feet long, it had a man-sized basket affixed to the bottom of its center pole, from which lines ran to the perimeter of the canopy.

The crowd was growing impatient; in an hour or so it would be time to go home for dinner. Finally, Garnerin got the balloon to inflate fully and scrambled into the parachute's basket. As it rose upright, they could see the arrangement: a rope ran from the bottom of the balloon down through a tube along the umbrella's pole and into Garnerin's basket, where it was secured. When the time came, Garnerin would cut the rope, which would release the balloon skyward and send him plummeting.

Garnerin waved as his craft cleared the dark green treetops and shrank slowly into the heavens. While the park faded into the gloom of shadows, the setting sun lit the balloon and umbrella with an orange glow, and after a long ascent, at two thousand feet above the city, it shone like a small golden moon floating against the darkening blue of the Parisian sky.

At least now they were seeing a balloon demonstration—not what was promised and not their money's worth, but it was something.

Then suddenly, the balloon exploded. Garnerin hurdled downward. After just enough time for a collective gasp, they saw the umbrella pop open, arresting his fall—but then the basket began to swing wildly. Whipping back and forth like a crazed pendulum, it threatened to catapult the hapless Garnerin right out of his craft. As it dropped, the crowd surged to follow its path, people breaking into a run toward where they guessed it would land, stumbling as they tried to keep their eyes on the man sailing across the sky. Garnerin, desperately gripping the sides of the heaving basket, could only catch glimpses of the mob racing along the streets below, as the land, from his gyrating perspective, roiled like a churning sea rushing up to meet him. Eighty feet, fifty, twenty, ten, two—and then a hard bump as the basket scraped and lurched to a stop. The umbrella collapsed, its pole flopping off to the side.

In minutes the crowd, out of breath and stunned by what they had seen, overwhelmed Garnerin as he struggled from the basket. He looked quite ill from the wild ride, but he was clearly uninjured.

All the doubts and ridicule were now swept away; women were weeping, some even fainting (according to reports), and everyone was shouting and cheering. Several men grabbed Garnerin and hoisted him to their shoulders. Carrying him high above the suddenly euphoric throng, they set him on a horse to carry their new hero back to the launching place of his glorious triumph.

THE IDEA OF a parachute, if not the name, had been around for centuries before Garnerin made his famous jump; after all, the principle on which it worked was obvious to anyone who had seen leaves float to the ground: flat, thin objects fell more slowly than solid ones. Hundreds of years before the birth of Christ, there were already folktales in China about people who had jumped from high towers or walls using wide-brimmed hats or parasols to slow their falls.

By the late fifteenth century, sketches of actual parachutes began to appear. One, attributed to an Italian engineer, circa 1485, shows a man hanging beneath a conical parachute. Around this same time, Leonardo da Vinci made a similar drawing in his *Codex Alanticus*, along with this description: "If a man has a tent of closely woven linen without any apertures, twelve *braccia* across and twelve in depth, he can throw himself down from any great height without injury."[2] Da Vinci's measurements, as it turns out, yield proportions roughly comparable with the size of a modern parachute.

A hundred years later a Venetian architect named Fausto Veranzio also depicted a man hanging from a parachute in his book *Machinae Novae* and was reported to have used the design to fall safely from buildings.

In 1688, the French envoy to Siam, Simon de la Loubère, reported seeing a tumbler, "supporting himself by two umbrellas, the [handles] of which were firmly fixed to his girdle,"[3] leap from a tower to entertain Siamese royalty. This account has the ring of truth, as Loubère noted that the tumbler often went out of control, landing in a river or tree, which kept the king mightily amused.

All, or none, of these early tales of people using parachutes or similar devices might be true; the accounts are simply too sketchy to confirm. It's also possible that Garnerin himself had been beaten to the punch by one of his rivals. Lenormand, the professor who named the parachute and is credited with being the first to drop animals with it in 1783, claimed to have also parachuted himself that same year—from the Montpellier observatory. Joseph Montgolfier, coinventor of the hot-air balloon, said he had jumped from the roof of his house. Blanchard, the showman who dropped animals from his balloon, claimed that in 1785 he had had trouble with his craft and had been forced into making a quick exit using a parachute—resulting in a broken leg, which he exhibited as proof.

But none of these men had witnesses, whereas Garnerin had hundreds. Thus, the effect of his feat was altogether different from what it would have been had he or any of his rivals carried out a demonstration just for the benefit of a few reliable observers—had Professor Lenormand, for example, parachuted before his colleagues at the Paris Conservatory. That would have been noted as a significant *scientific* achievement. Instead, Garnerin's stunt would come to have far more impact on popular culture than it did on science. In the years after his jump—for a full century—there would be almost no scientific demonstrations or attempts to improve the new invention. Instead, scores of parachutists would tour Europe and eventually America, dropping from their balloons strictly to thrill paying customers.

## Dancing on Ladders

To understand why parachuting had such an impact on culture, we should ask how people felt about falling before the changes of the 1700s. The short answer is, they were terrified of it.

The idea of intentionally falling or risking a fall from a great height for recreation or to entertain had for aeons been almost completely inconceivable—it simply made no sense. The closest the Greeks, Romans, Egyptians, and Chinese got to falling as a regular activity was in their traditions of acrobatics, activities depicted on the pots, walls, and tombs of the ancient world. On Greek and Roman artifacts we see people leaping from horses, shinnying up tall poles, and turning somersaults. There are also accounts in all these cultures of entertainers performing simple gravity tricks, such as springboard leaping, as part of acts that included juggling, dancing, and singing. Also common, at least in the cities, were performers who would dance along a stretched rope. There's no way to tell how old this trick is; pictures of it appear in cultures all over the civilized world, from the Mediterranean to China.

Yet none of these ancient acrobatics involved a risk of falling very far. Rope dancing, for example, wasn't the sort of dangerous high-wire act we think of today; it was a feat of skill, usually performed only a few feet off the ground. It certainly had more to do with dancing than with daring. In sum, acrobats before the eighteenth century had little interest in risking their lives for show.

There are only a handful of recorded instances in which people living before the modern era regularly took dangerous falls on purpose, the land diving of the Pentecost Islanders being one example. Another is a ceremony practiced in the Americas among the Aztec and Hopi. As with the Pentecost Islanders, it wasn't meant to be fun. It was instead a ritualized test of manhood, imbued with religious meaning, further suggesting that falling was regarded with a cosmic dread far too frightening to be indulged in for entertainment or recreation.

The Aztec version, called Los Valadores, is a ritual at least fifteen hundred years old. It's still performed in Mexico today (having now *become* entertainment, at least for visitors) in San Luis Potosí, Papantla, and Huasteca, where it's performed by the Aztecs' descendents.

The rite of Los Valadores (the flyers) begins in the spring, with villagers journeying to the forest to select a tall, sturdy tree, perhaps one hundred feet high; this will become *el volador*, the flying pole. Once selected, they ask the tree for its permission to be used in the sacred ceremony; only then is it blessed and felled. The tree is dragged overland to the village by hundreds of tribesmen, where it is stripped of its branches to create a tall, straight pole. Four limbs are then cut to equal lengths and lashed together to form a square, which is suspended from its corners at the top of the pole in such a way that it can rotate, like a pinwheel. Four long ropes are wrapped around the top few feet of the pole, their ends secured to the top. Finally, the entire pole is hoisted upright in the village center.

On the day of the ritual, five men, dressed in bright costumes and feathers, ascend. The group's leader lies down on the hub of the rotating square. The other four men sit precariously on the bars that form the square's edges, each taking an end of one of the four ropes wound around the pole and tying it to his ankles. As flute players and drummers perform a hundred feet below, the men simultaneously drop backward from their perches, heads down, arms outstretched. Their weight pulls the rope from the pole, and as it unwinds, the square frame begins to turn, swinging the men out in great circles, like a carnival ride, dropping them slowly toward the ground.

The ritual is still a religious one, although the specific meaning it had to the Aztecs is open to speculation. But in it, falling is clearly seen as a terrifying act meant to show the courage of the young men who try it. And while today's version doesn't really involve much risk (many of the performances are done on stable metal structures), there is reason to think it wasn't always so safe. Different versions of the ritual were widespread across Mesoamerica, and one formerly practiced by the Hopi of the southwestern United States was far more frightening than the gentle drop of Los Valadores.

The Hopi ritual, now extinct, was called Sáqtiva (ladder dance). It was performed on a narrow ledge 150 feet above the canyon floor near the now abandoned village of Pivánhonkyapi that once stood on one of the Hopi mesas in Arizona; here, you can still see the dish-sized holes carved into the rock that held the ceremony's four poles.

As in the Aztec ritual, the trees used were found in a distant forest, carefully selected and blessed, and then brought to the village. Two trees were stripped bare and wound with leather thongs that spiraled down their lengths; two others were left with fragments of branches jutting out from the trunk—just enough to allow for climbing. On top of each pole a crossbar was attached. The four poles were then planted into the holes in the narrow rock ledge.

At the beginning of the ritual, chanting tribesmen and musicians playing drums and flutes would line the high wall behind the ledge. Four naked young men then appeared; each painted for death, because death was a real possibility. They took their places: two ascending to the tops of the climbing poles to face each other; the other two standing at the base of the wall, a few paces from the thong-wrapped poles. Then they waited for the signal.

When it came, all four sprang into action. The two men at the top of the poles would simultaneously leap across the gap, passing each other in midair, to grab the crossbar of the opposite pole, swinging their bodies out over the canyon. At the same moment, the two other young men would leap from their stances as if to sail off the ledge, but would instead grab the leather thongs and begin to spiral outward around the pole and over the drop, like human tetherballs, as the thongs unwound. If any of the men lost their grip, the fall would be fatal.

According to the Hopi's oral history, one season brought a double disaster that ended the ritual. By tradition, the young men who were chosen as fliers were required to be single, without a wife or any other romantic involvement, to ensure that there would be no grieving mate if a man plummeted into the canyon. But this particular year one of the men had been pursued by a woman whom he had spurned. The legend says that three times the rejected woman cut through the pole he was to fly from, and each time Másaw, the god of the underworld, mended the break. Finally, on the night before the dance, she cut through the log again, and when the dancer grabbed the thong in his leap, the pole broke, sending him hurtling to his death. When the village elders later gathered in the village's ceremonial chamber, the *kiva*, to discuss the bad omen, the roof collapsed, killing them. The gods' message was clear: the ritual, and the village, were immediately abandoned.[4]

IN THE AMERICAS, the practice of risking a high fall as a public event never grew beyond these religious ceremonies. But in Europe, the minstrels—who typically weren't acrobats per se but generalists who would sing, dance, and juggle, as well as tumble and rope-dance—began to specialize a little more, and sometimes (usually for a special event such as a coronation) they would try something truly dangerous. By the late Middle Ages, we find the first reports of minstrels risking serious falls to entertain the public.

At the wedding of Charles VI and Isabel of Bavaria in Paris in 1385, one performer walked along a rope stretched from the highest house on the bridge of Saint Michael to the highest tower of Notre Dame, singing and holding burning candles along the way. At the coronation of King Edward VI in 1547, a Spaniard slid on a board headfirst down a steeply stretched rope from the battlements of St. Paul's Cathedral to the ground (which King Edward probably enjoyed a great deal, since he was nine years old at the time). An engraving from the seventeenth century shows a crowded Nuremberg courtyard bounded by galleries four stories high, with tightropes stretched between them and tightrope walkers holding balancing poles working their way up and down the angled ropes.

These life-threatening stunts, though, were rare. If antiquity had been filled with even half the daredevils that appeared in the eighteenth century, certainly many more than the few accounts we have would have survived. The picture we are left with is that throughout history and into the Renaissance, people not only avoided falling themselves but also almost never saw anyone else take a high fall or risk one. That changed, and changed dramatically, in the eighteenth century.

## Gravity Shows

The shift began as the century dawned. As population and wealth in Europe grew, the wandering minstrels began to organize into traveling fairs and eventually settled down into more or less permanent exhibitions. In Paris there were the St.-Germain, St.-Laurent, and St.-Ovide fairs; in London, there were fairs at Bartholomew, Southwark, and Sadler's Wells, just outside the city.

With larger, regular audiences to entertain, competition among gravity performers suddenly accelerated. Springboard leapers vied for the highest and longest jump. One, holding flaming torches in each hand, catapulted over an orchestra in 1697; another flew over fourteen people in 1727, turning a somersault as he went. In England, in 1741, Nicolini Grimaldi bounced from the stage and sailed straight out over the audience, grabbing on to a high dangling chandelier at the last moment, which sent a hail of crystal raining down on the Turkish ambassador, who happened to be sitting in the front row. In France, adding the possibility of a particularly gruesome end, a leaper flew over twenty-four men with up-pointed swords.

Gravity performers were gradually becoming stars. No longer anonymous street acrobats only a notch above panhandlers—as they had been for centuries—they were developing international reputations. In 1704, an advertisement in the British newspaper *Postman* touted the fame of one event's performers:

> During the time of May-Fair, will be seen a compleat Company of near 10 of the best Rope-Dancers, Vaulters and Tumblers in Europe, who are all excellent in their several performances, and do such wonderful and surprizing things as the whole world cannot parallel; where Finley, who gave that extraordinary satisfaction before Charles III King of Spain on board the *Royal Katherine*, performs several new entertainments, and where the Lady Mary, likewise shews such additions to her former admirable perfections, as render her the wonder of the whole world.[5]

Rope dancers began to do more than just dance. They began to balance in boots, wooden shoes, even with baskets or chains on their feet. They balanced on slack ropes, walked along sloping ropes, and vaulted on high ropes. In 1754, it was advertised that Mons. Vangable would "jump over a garter his own height forwards and backwards,"[6] but he was outdone that same year by Gagneur, who turned a back flip on a rope stretched seven feet off the stage.

The skill that had been the stock in trade of gravity performers for centuries was giving way to a new kind of thrill, the thrill of seeing someone risking a serious fall. In 1732, a flyer—an acrobat who slid down a slanted rope strung from a steeple—hit his knee on a chimney and fell, "whereby," the *Grub-street Chronicle* reported, "he broke his wrist, and bruised his head and body in such a desperate manner that 'tis thought he cannot recover. — On Saturday he died."[7] In 1750, the Turkish rope dancer Mahommed Caratha introduced walking on a wire, which was far thinner and so more dangerous than rope. In 1753, at Bartholomew Fair, a wirewalker fell and was seriously hurt.

For the first half of the century, dangerous gravity performances were largely limited to the fairs in the big cities. But then, in 1768, Philip Astley invented the institution that would bring them to the larger world: the circus. The circus began as a display of horsemanship; Astley, a sergeant major in the British army, would gallop his horse in a tight circle while he stood on its back, using centrifugal force to keep from falling—thus, the circus ring. But he soon added wirewalkers, tumblers, and acrobats, and as the shows spread (Astley founded nineteen circuses in Europe, and by the 1790s, circuses were showing up in Russia and America), gravity performers began to take center stage. Eventually, death-defying gravity stunts would become the headliners, while the clowns, animals, ground acrobats, and other performers would become opening acts for wirewalkers and, in later years, trapeze artists, high divers, and human cannonballs. With the rise of the circus, the image of a human risking a deadly fall would become familiar to far more people than ever before.

## Inventing Mountaineering

But the falls risked by circus performers, perilous though they were, would seem like a stumble off the front porch compared to the unspeakable falls humans would soon risk in another eighteenth-century invention: mountaineering.

It may seem odd to say that mountaineering was "invented," but as an activity for its own reward, it's fair to say that it was. Before its invention, the human attitude toward mountains was simple: avoid them. They were avoided for the very practical reason that they were dangerous, being both cold and a good place to fall. And they were avoided because in many ancient cultures they were the home of gods, and gods got nasty when disturbed.

Well into the Middle Ages, when people were less inclined to think gods actually lived among the peaks, they still believed that witches, demons, and malevolent spirits dwelled there. The Alps in particular—

where mountaineering began—had a bad reputation. In the fourteenth century, the people of Lucerne believed that the body of Pontius Pilate (after having been moved over several centuries to different locations) had wound up in a lake on nearby Mount Pilatus. They also knew that his ghost arose annually on Good Friday, and if you saw him, you would die within the year. He also caused blizzards when he was upset, which happened when people threw stones into his lake. That being the case, the authorities forbade anyone to climb the mountain without an official guide. When, in 1387, five clerics tried, they were thrown in jail.

The belief that awful beings lived in the mountains persisted until the dawn of the Enlightenment. Travelers reported finding a race of subhumans in Alpine valleys. (In this there was a grain of truth: inbreeding and a lack of iodine, which causes cretinism and goiters, had left some villages filled with deformed, mentally impaired people.)

What's more, there were dragons. As late as 1723, there existed scholarly work (parts of which were reprinted in a standard reference on the Alps in 1730) that took the existence of dragons in the Alps as an established fact. This was not the word of a superstitious crackpot but of Johann Jakob Scheuchzer, professor of physics at Zürich University and a fellow of the same Royal Society of London to which Sir Isaac Newton belonged. Scheuchzer traveled extensively in the Alps between 1702 and 1711 for his research and produced a respected volume discussing everything from fossils to glaciers. And when it came to collecting information for his chapter on dragons, he was no fool: he knew peasants told wild stories, so he was careful to talk only to eyewitnesses, making sure to get all the details of time, place, and circumstances. Even then, he would look for corroboration from villagers of standing, such as officers and pastors. Here is the account he got from Christophe Scheurer, the first magistrate of the region:

> On one particular night, when I was contemplating the serenity of the sky, I saw a brilliant dragon take to flight from a corner of the great

> rock of Mont Pilate; his wings moved with great speed. His torso was long as was his tail and neck. His head was that of a toothed serpent. As he flew, sparks, such as those that fly from a hot iron hit against an anvil by the blacksmith, flew from his body.[8]

Scheuchzer collected dozens of these accounts, most of which he dismissed, which made his report all the more credible. In the end, he cataloged the different species of dragons, noting their habitats (dragons of Lucerne, for example, being different from those of Zurich) and meticulously differentiating between those that had a cat's head on the body of a snake, for instance, and those that had bat's wings.

By the mid-1700s, though, a few scientists began to ignore the superstitions, and soon locals who lived in the shadow of the Alps began guiding naturalists and geologists up to the glaciers and lower summits. Then, in 1760, naturalist Horace-Bénédict de Saussure sparked the event that would burst the height barrier: he offered a monetary prize (the amount is unknown) to the first person to climb Mont Blanc, which at 15,782 feet was Western Europe's tallest peak.

For twenty-five years, various attempts were made on the mountain. Saussure's guide, Pierre Simon, tried twice soon after the reward was offered, but failed. Marc-Théodore Bourrit, a choir leader at Geneva Cathedral and self-proclaimed "Historian of the Alps," proposed what he thought were the best routes, but the local guides who tried still couldn't summit. In 1783, Bourrit teamed up with Michel-Gabriel Paccard, a local doctor and amateur scientist, but they couldn't reach the top either. The next year, each of them tried again, separately, with the same result. In 1785, Saussure himself tried, accompanied by Bourrit; again, failure.

It wasn't that Mont Blanc was too steep or technically difficult to climb. What stopped the men was the simple fact that because no one climbed high mountains, no one really knew how. And if anything went slightly wrong, they had no idea how to cope. Typically, they were weighed down with far too many supplies; Saussure and Bourrit took seventeen porters to carry all the blankets, pillows, food, and wine they thought they would need for what would be, if successful, a three- or four-day trip.

Moreover, their routes and decisions were based on the style of hiking they had always done, which didn't work very well when applied to the heights. For instance, instead of aiming for rock, which later mountaineers would discover offered routes that were not only less fatiguing but also safer (because you were less likely to be carried away by an avalanche), these early climbers preferred the softer, more familiar feel of snow. If the snow became too deep, they would turn back. Accustomed to abandoning a hike if the weather turned bad, they were unprepared, physically and psychologically, to deal with a constant of high mountains—weather that turned bad suddenly and often—and so they would turn back. And as the air thinned alarmingly—more than they were used to because no one had climbed as high before—they would become exhausted. And they would turn back.

Sheer determination was what was needed, and sheer determination is exactly what Jacques Balmat, a twenty-five-year-old local guide, had plenty of. After deciding, in the summer of 1786, that he would claim Saussure's prize, Balmat vowed to himself that what had stopped the others wouldn't stop him. He wouldn't make the mistake of mounting a bloated expedition; he would travel alone and light. On June 5 of that year, he set out, unencumbered by porters or anything else. He took with him a loaf of bread and a bottle of brandy. He didn't even take a blanket.

Balmat had his route planned, and all went well until he reached the Grand Mulets, a rock outcropping above which thick fog had moved in. It didn't stop him. In spite of the fact that he was alone and had no shelter, he decided to stay through the night. All night long he stamped his feet and clapped his hands together to stay awake.

The next day he kept climbing, and when night fell, he again stamped and clapped himself awake until dawn. Although determined, Balmat wasn't foolhardy, and so the next morning, after surviving two nights on the mountain, he concluded that he had better not push his luck, and so he began to descend. On the way down, he met three guides on their way up who asked him to join their ascent. Exhausted as he was, he couldn't bear the thought of being excluded, so he agreed. But first—incredibly—he descended all the way back to his home to tell his wife he would be longer than expected. He took some food and fairly *ran* back up.

Rejoining the party at 11:00 P.M., he climbed with them through the third night. The next morning, two more guides joined them, making a group of five besides himself. He couldn't stand the idea of sharing the prize with so many, and so he struck out on his own. Later, splayed out on a ridge which "seemed a path fit only for a rope dancer," he looked down. And there, he saw, was a much better route.

He descended and discovered that his competitors had left. With what must have been a superhuman effort, he seized the opportunity and spent the rest of the day exploring the new route and spent his fourth night on the mountain. At dawn, again knowing his own limits—which he had already proved to himself were greater than anyone else who had attempted the climb—he retreated.

He finally stumbled home later that day, sunburned and defeated—but only temporarily. With his new route in mind and realizing he could use at least some help, he approached Paccard, the doctor who had made the 1783 attempt with Bourrit. Paccard wanted to take two guides, but Balmat refused, sticking to his instinct to travel as lightly as possible.

Balmat and Paccard set out together on August 9, 1786. According to Balmat, at practically every step Paccard trembled with the fear that had plagued others who had challenged the mountain. A rising wind forced them to crawl, and Paccard was petrified—but Balmat urged him on. When they reached the Dôme du Goûter, where the air was exceedingly thin, Paccard refused to budge. Balmat made the final push alone. Fighting tremendous fatigue ("I had to stop almost every ten steps and wheeze like one with consumption,"[9] he later reported), Balmat staggered to the rounded dome that was the summit of Mont Blanc.

"I no longer had any strength to go higher," he said, "the muscles of my legs seemed only held together by my trousers. But behold I was at the end of my journey; I was on a spot where no living being had ever been before, no eagle or even a chamois! I was the monarch of Mont Blanc! I was the statue on this unique pedestal."[10]

By the time they began their descent, Paccard had become completely snowblind (another hazard early climbers were unprepared for) and had to be led by Balmat, but they arrived home safely—and Balmat collected Saussure's reward.

Locally, Balmat was lionized; he became the region's top guide, and a statue was erected in his honor. But on the general public, Balmat's accomplishment hardly registered. So a guide had climbed higher than before—what of it? That didn't change the fact that mountains were filled with horrors. The only people interested in mountain climbing were scientists, and since Balmat and Paccard had added nothing to scientific knowledge, scientists ignored them. The next year, Saussure, who was a respected naturalist, summited Mont Blanc, making careful measurements and observations along the way. In the scientific community, Saussure's climb was hailed as a landmark achievement.

Yet—although no one could see it at the time—like Garnerin with his parachute drop, Balmat had tapped a new wellspring of gravity's allure. That allure was implicit in how Balmat had described his feelings at the summit. "I was the monarch of Mont Blanc!" he had declared. "I was the statue on this unique pedestal!" It was a new sensation, one that came from fighting gravity all the way to the top of a high peak. Scientists may have ignored it, but it was the battle against gravity that would come to define mountain climbing, a confrontation that would be refined and purified so that eventually what mountain climbers would come to value would be the *quality* of the fight on the way up. These new adventurers—soon to be called "mountaineers"—would, in the ensuing years, take the common wisdom that had kept people out of the mountains since the dawn of time and turn it on its head. Instead of avoiding the highest and steepest peaks, or at least looking for the safest route up, they would aim directly for mountains and routes that would guarantee the greatest possible chance of plummeting. It was this goal—to maximize the risk of falling—that would come to define modern climbing.

## Mechanical Mountains

The allure of gravity would also, in the eighteenth century, seduce Europeans in another form, this one directly and immediately accessible to the public: the roller coaster. Sliding down a slope wasn't a new sensation; roller coasters, in fact, grew from tobogganing—or rather from a kind of tobogganing that had already been designed to heighten the sensation of falling. Since the sixteenth century, peasants in the backwoods of Russia had been building artificial slopes more efficient for sliding than a natural mountainside. They would raise a wooden frame, some seventy feet high, and fill it with snow, packing it down and tapering out the end into something like a toboggan run. But then they used two tricks that turned the slope into a more effective falling machine. First, they watered down the snow to freeze it into a solid trough of ice. Next, instead of using a toboggan, they nestled into a block of ice with a carved seat stuffed with a little straw—essentially a frictionless sled—that shot down the slide so fast that the runout had to be covered with sand to stop a disastrous end-over-end tumble.

Although ice slides had been around for three hundred years, it was in the eighteenth century that they caught the public's attention. In St. Petersburg, during the winter festivals when the city's network of rivers froze, crowds would gather to ride ice slides, now erected with spacious stairways to the top and festooned with flags and decorations. And it was in 1784, also in St. Petersburg, that the ice slide was freed from the limitations of ice and transformed into a machine that could deliver the thrill of falling anywhere, in any season. That year, the "Russian Mountain" was re-created in mechanical form, with wheeled cars on a track. (Today, in Europe, a roller coaster is often still called a Russian Mountain.) Because St. Petersburg was unusually cosmopolitan at the time—a seaport whose population was one-quarter sailors and soldiers—word of the new invention spread fast. Twenty years later, a similar ride appeared in Paris, this one employing sleds with runners that coasted over tracks with rollers—the "roller coaster."

After its invention, the roller coaster became faster and scarier in short order. In 1817 Parisians built the Promenades Aeriennes, which had cars that rocketed along at forty miles an hour, a startling speed for a society used to the trot of horse-drawn carriages. By 1846, the French abandoned nature's model completely and built something more frightening than any mountain: the first looping roller coaster, which turned riders completely upside down. Because the loop was small—only thirteen feet—its centrifugal force was severe; it shoved riders into their seats with such spine-wrenching force that few people could bear to ride it, and it was eventually demolished.

Still, the future had arrived. The roller coaster was more than the first machine to deliver the sensation of falling to the masses; it was a machine that *amplified* the sensation of falling. The drops were steeper and faster than those from tobogganing—and you couldn't choose a gentler route or drag your feet to slow the thing down. By stripping away control, increasing the magnitude of the falls, and distilling it all into a few minutes, the roller coaster isolated, packaged, and fetishized the sensation. Like sugar refined to be so sweet that even the natural sweetness in fruit pales by comparison, roller coasters would refine—and redefine—what it felt like to fall. And as the ensuing decades would prove, once people tasted it, they couldn't get enough.

## The Gravity Hero

With the circus, mountain climbing, the roller coaster, and parachuting, Europeans completely overturned what falling had meant to previous generations in a single century: risking a fall was now big entertainment delivered by circus performers, a new alpine sport indulged in by the wealthy, and falling itself a sensation consumed by adults on amusement rides for fun.

It's hard to know what accounts for the change. New technology helped in two cases: ballooning made parachuting easier than it might have been from a tower or cliff, and improved metallurgy probably made building a roller coaster more practical. But neither of these were decisive, and there were no new tools or materials that encouraged mountaineering or the popularity of circus daredevils. It's possible that the rising population and growing wealth of the time simply accelerated what people would have done anyway, as an expanding leisure class might be expected to make improvements in all kinds of recreation and entertainment. Still, the shift seems too pronounced for this simple explanation.

Flinging yourself down a roller coaster or watching a man risk his life on a high wire, we might guess, was somehow consistent with a new perspective on life and the world—and we don't have to look far to find a new perspective.

Eighteenth-century Europe saw the dawn of the romantic movement, one that swelled to help create the French Revolution in 1789 and one that glorified heroes, in particular, heroes who arose from the masses to face seemingly insurmountable odds. It may be that wirewalkers and parachutists appealed to this new taste for courageous rebels. And because gravity heroes weren't political, they could be admired without subscribing to dangerous ideas.

Aside from its metaphorical content, the act of defying gravity was potent in itself because it confirmed the Enlightenment's version of progress; of man as nature's master, not her victim. Defeating gravity was the most dramatic conquest of physical law people had ever seen, largely because they could actually see it. While the eighteenth century was filled with advancements in mathematics, physics, astronomy, hydraulics, optics, and medicine—any of which, it could be argued, were more significant in demonstrating man's power over nature than parachuting—these were incremental advances carried out over decades. In contrast, every gravity performance told its story in a few minutes and within a dramatic three-act structure: a wirewalker steps on to the wire, struggles across, and arrives safely—or falls. A parachutist leaves his balloon, battles the wind through the fall, and then lands safely—or dies. What was more, on roller coasters, people could live through the drama of their own conquest of gravity—they could, at least temporarily, become gravity heroes themselves.

And these new heroes may have also represented something more. Like gravity, class oppression in the eighteenth century was felt as an immutable force of nature that had held people down for as long as memory had existed. To witness Garnerin float up to the clouds and then drift safely down must have been an exhilarating demonstration that limits once thought permanent could be overcome. If the space between heaven and earth could be navigated, then perhaps the hierarchies of class could be traversed as well.

As if to confirm that gravity performers were indeed iconoclasts, voices of authority attacked them and their stunts from the beginning. As early as 1574, a troupe playing in London was denounced by the Recorder of London for "the unchaste, shameless, and unnatural tumbling of the Italian women."[11] In 1709, Richard Steele, editor of two influential British dailies, the *Tatler* and *Spectator*, criticized one showman who dared to substitute vulgar gravity performers for the plays Steele thought provided more elevated entertainment, saying the showman, "having no understanding in his polite way, brought in upon us, to get in his money, ladder-dancers, rope-dancers, jugglers, and mountebanks, to strut in the place of Shakespeare's heroes and Jonson's humorists."[12] In 1711, Steele suggested in a *Spectator* commentary that since "bodily actors" appealed only to those who couldn't appreciate anything better, they should be separated from legitimate actors:

> The Method, some time ago, was to entertain that Part of the Audience, who have no Faculty above Eyesight, with Rope-dancers and Tumblers; which was a way discreet enough, because it prevented Confusion, and distinguished such as could show all the Postures which the Body is capable of, from those who were to represent all the Passions to which the Mind is subject. But tho' this was prudently settled, Corporeal and Intellectual Actors ought to be kept at a still wider Distance than to appear on the same Stage at all: For which Reason I must propose some Methods for the Improvement of the Bear-Garden, by dismissing all Bodily Actors to that Quarter.[13]

That same year, reader Ralph Crotchet wrote to the *Spectator* to defend gravity performers, claiming that watching them could help audience members become more nimble:

> Every Man that goes to a Play is not obliged to have either Wit or Understanding; and I insist upon it, that all who go there should see something which may improve them in a Way of which they are capable. In short, Sir, I would have something done as well as said on the Stage. A Man may have an active Body, though he has not a quick Conception; for the Imitation therefore of such as are, as I may so speak, corporeal Wits or nimble Fellows, I would fain ask any of the present Mismanagers, Why should not Rope-dancers, Vaulters, Tumblers, Ladder-walkers, and Posture-makers appear again on our Stage? After such a Representation, a Five-bar Gate would be leaped with a better Grace next Time any of the Audience went a Hunting.[14]

These were just the opening salvos in what would become the battle over gravity play that continues to this day. Rope dancers and tumblers would be the first to be attacked, but parachutists, high divers, trapeze artists, mountaineers, even roller-coaster designers—in fact, just about anyone involved in a new form of gravity play—would eventually find themselves under fire.

The image of the gravity hero, though, wouldn't go away. Instead, that image—a rebel who rises from the masses to confront nature on behalf of the common man—became a powerful and enduring archetype. Since its inception in the eighteenth century, it has shaped the view of virtually every gravity challenger since. It is with us today, updated in our cynical age as antiheroism and embodied in rebellious skateboarders, pot-smoking Olympic snowboarders, and outlaw BASE jumpers. It is with us in nearly every portrayal of an extreme athlete we see.

Sam Patch, the "Jersey Jumper," so enthralled nineteenth-century America with his hundred-foot plummets that Nathaniel Hawthorne devoted an essay to him, Edgar Allen Poe critiqued a poem about him, and national magazines from *Harper's* to *The Century* either praised or condemned him. Patch is pictured here leaping from the edge of Genesee Falls in a contemporary illustration, circa 1830. *(Courtesy Rochester Public Library)*

CHAPTER 2

# Terrified Imagination

It has been a common thing for persons to leap from the top of the large stone six-story factory, below the bridge into the whirling and boiling current eighty feet below, and we have never heard of any accident has yet happened. —*Rochester Daily Advertiser*, circa 1828

In 1835, Nathaniel Hawthorne traveled to Rochester, New York, drawn by the legends that surrounded Sam Patch, known as the "Jersey Jumper," a man who would become America's first major gravity hero. Born in Pawtucket, Rhode Island, in 1807, Patch was a cotton spinner. When he was just out of his teens, his business partner took off with the money from their New Jersey mill, leaving him broke. How he got the idea to jump for money isn't recorded, nor is the number of jumps he made, if any, before the one that made his name.

On the day of that jump, September 30, 1827, a preassembled bridge was being moved over Passaic Falls in New Jersey, and a crowd had gathered to watch. Presented with a convenient audience, Patch announced that he would jump from a ledge on the cliff eighty feet above the water, and collected money from everyone willing to make a donation. When he had gotten enough, he climbed to the ledge and stripped to his undershirt and shorts. He threw a few stones over first, estimating how far he would have to jump to clear the outcropping of rock below. Then he backed up, took a short run, and leapt. His body tilted a bit in the air, but he entered the water cleanly and bobbed up to the cheers of the crowd on the bank. The former cotton spinner suddenly had a new, profitable career.

Over the next two years, Patch jumped three more times from Passaic Falls, as well as from other cliffs and from ships' yardarms, working his way up to one-hundred-foot drops. His audiences steadily grew, while reports of his feats appeared in numerous national publications. Patch often stoked the publicity by taking out advertisements.

In 1829, Patch's growing celebrity attracted the attention of a group of businessmen who were looking to attract tourists to their resort in Upstate New York. Their offer: to pay the Jersey Jumper handsomely if he would leap into the waters of Niagara Falls.

Patch was to headline a bizarre series of events that characterized the fringes of nineteenth-century entertainment. First, the hotel owners would use gunpowder to blow up huge chunks of the cliffs surrounding Niagara Falls, purely for the dramatic effect. Next, they would send a schooner filled with animals crashing over the Falls. Two years earlier, these attractions had jammed the shores with fifty thousand spectators. Now they would repeat the formula but cap the show with Patch, a human challenging the Falls for the first time.

On October 7, 1829, as the extravaganza began, the cliffs did explode, but the most spectacular demolition was called off because officials feared it would change the course of the river. The ship filled with animals got caught on a rock; the audience watched in disappointment as most of the bears, cattle, and other assorted creatures simply jumped off and swam to shore. It was left for Patch to hold the dwindling crowds, and things didn't go well. It had begun to rain, and a chain holding the enormous ladder he was to climb broke, forcing people to endure the downpour for hours until it could be fixed.

Finally, at four o'clock, Patch climbed the ladder that leaned out from the cliff at a steep angle, supported at its base at the bottom of Goat Island, the rock edifice between Horseshoe and American Falls. The trappings of heroism were everywhere: "Flags (were) streaming, bearing mottoes full of liberty and patriotism,"[1] wrote one witness. Patch stood on the tiny platform, eighty-five feet over the river, blew a kiss to the ladies, and kissed an American flag. Then he dove headfirst.

He disappeared into the churning water. A boat circled the spot to pick him up, but he was gone. When the worried crowd finally caught sight of him climbing out onto the riverbank, they exploded in cheers.

A few days later Patch repeated his Niagara dive, this time from 130 feet. One witness who saw him captured the strange mix of emotions sparked from watching this kind of stunt. The writer focused on the moment of seeing that Patch wasn't dead. "Then it was," he wrote, "that a painful and unpleasant, yet indescribable sensation was driven from each breast, by the flood of joy which succeeded, on seeing he was safe."[2]

After his Niagara triumph, Patch traveled to Genesee. Perhaps picking up a tip from the Niagara showmen, Patch spiced up his performance by first shoving a bear off the hundred-foot cliff into the river. After the bear swam to shore, Patch jumped, slicing into the water perfectly.

A week later, he planned to jump again at the same spot, but this time he would extend the challenge of the natural cliff by building a twenty-five-foot platform on top. "Higher Yet! Sam's last jump!"[3] the posters roared; the implication seemed to be that after this jump, he would retire.

On November 13, 1829—Friday the thirteenth, it so happened—Patch climbed up to the platform. He took a strong drink—the air was cold—and then turned to the several thousand people who had gathered. He delivered a brief speech in which he compared his accomplishments to military victories and himself to Napoleon. Then he waved and stepped off.

Witnesses said he started to tilt almost as soon as he left the stand. In midair, he broke from his usual ramrod posture and began to flail, tipping over until he was nearly parallel with the water. He hit with a loud smack. Minutes passed, and he didn't surface—in fact, he never surfaced that day.

## Trifling with God

Nathaniel Hawthorne had not seen any of Patch's jumps; he had only heard the stories. Six years after Patch's disappearance, Hawthorne went to Genesee Falls to reflect on what it all meant. In 1835, his thoughts were published in the *New-England Magazine*:

> My chief interest [in Genesee Falls] arose from a legend, connected with these falls, which will become poetical in the lapse of years, and was already so to me, as I pictured the catastrophe out of dusk and solitude. It was from a platform, raised over the naked island of the cliff, in the middle of the cataract, that Sam Patch took his last leap, and alighted in the other world. Strange as it may appear—that any uncertainty should rest upon his fate, which was consummated in the sight of thousands—many will tell you that the illustrious Patch concealed himself in a cave under the falls, and has continued to enjoy posthumous renown, without foregoing the comforts of this present life. But the poor fellow prized the shout of the multitude too much not to have claimed it at the instant, had he survived. He will not be seen again, unless his ghost, in such a twilight as when I was there, should emerge from the foam, and vanish among the shadows that fall from cliff to cliff. How stern a moral may be drawn from the story of poor Sam Patch! Why do we call him a madman or a fool, when he has left his memory around the falls of the Genesee, more permanently than if the letters of his name had been hewn into the forehead of the precipice? Was the leaper of cataracts more mad or foolish than other men who throw away life, or misspend it in pursuit of empty fame, and seldom so triumphantly as he? That which he won is as invaluable as any, except the unsought glory, spreading, like the rich perfume of richer fruit, from virtuous and useful deeds.[4]

Hawthorne's praise seems odd from our perspective, considering that if Patch were performing today he would most likely be viewed as just another crackpot desperate for fame. If reporters wrote about him at all, it would be with a smirk. Certainly no media outlet today would publish what the *Buffalo Republican* did: "The jump of Patch," it declared, "is the greatest feat of the kind ever effected by man."[5]

Yet Patch's celebrity was so great that for decades his name appeared in scores of publications. As with other gravity heroes, the reaction was both adulation and condemnation. In 1840, *Knickerbocker Magazine* published a lengthy ode to Patch entitled "The Great Descender," by Flaccus (a pseudonym of Thomas Ward), which envisions him sitting on a precipice and suddenly realizing that like meteors, lightning, men who drop from parachutes, and even the tower of Pisa, he can achieve fame by falling:

*From his drenched hat the rain-drops, gathering slow,*
*Drip one by one far down the gulf below;*
*Just then, a sudden meteor, trembling there,*
*Slid down the sky, and quenched itself in air.*
*The hero started; "Ha! I will obey!*
*Renown is mine!—the heavens have marked the way.*
*Yon meteor tells me, Wherefore climb at all,*
*Since fame as well irradiates things that fall?*
*Yon earth-born meteor, spawn of slime and mire,*
*More wakes the vision by its dropping fire,*
*Than the world-sprinkled heavens, whose lights sublime,*
*Have cheered the darkness since the birth of time.*
*And more; does not the monarch of the skies*
*Go down in glory too, as well as rise?*
*How many watch him as he sinks away!*
*How few pay homage to his rising ray!*
*The lightning's self may glitter as it likes,*
*'T is ne'er gazetted, save it stoops and strikes.*
*How many, smitten with the fame it gave,*

*Have dived in bells far 'neath the ocean-wave!*
*Or from balloons in parachutes gone down,*
*Stooping to catch the jewel of renown.*
*We pass unpraised the stones that round us lie,*
*But hail them when they tumble from the sky.*
*The Arch-Fiend's fame no poets tongue could tell,*
*Nor history chronicle, until he fell*
*And Pisa's tower, so bending, and so tall,*
*We laud, that only makes a threat to fall.*[6]

The poem was reviewed favorably in the *North American Review* and unfavorably by Edgar Allan Poe, who called it "elaborate doggerel"[7] and who couldn't understand how Ward could praise "the advantages to *Science* which have accrued from a man's making a frog of himself."[8] Poe also mentioned Patch in a humorous piece about "diddling," in which he mocks small-minded self-serving of every kind. Great works are to diddling, he says, "as a Mastodon to a mouse—as the tail of a comet to that of a pig—as Homer to Flaccus—as the *Iliad* to *Sam Patch*."[9]

Although Poe ridicules Patch, he didn't have to explain who he was because everyone knew. In a book on travel in Egypt published in 1838, the author belittles the region's waterfalls as inadequate by mentioning the famous leaper: "The great and ever-to-be-lamented Sam Patch would have made the Nubians stare, and shown them that some folks can do things as well as other folks; and we question, if there is a cataract on the Nile, at which that daring diver would not have turned up his nose in scorn."[10]

In 1850, when the *New Englander and Yale Review* wanted to criticize Ralph Waldo Emerson, they called him an "intellectual Sam Patch, who makes it his profession to go about the world, leaping down precipices, plunging into abysses, in every deep seeking a still lower [one] in which to expose himself for the sake of the applause and the pay, which men are always willing to bestow on any one who is fool-hardy enough to entertain them with such extravagances."[11]

In an 1866 issue of *Harper's* magazine, John Hay describes mentioning Patch's name to a Kentuckian he meets in France. "You have just pronounced," the man says, "the greatest name of our times. I hope you share my admiration for Patch!" The man goes on at length to claim that Patch had meant to commit suicide all along, but that he was admirable because he went out in a blaze of glory. In 1870, *Punchinello* published a list of the "Last Words of Eminent Men" and included Patch's (supposedly, "Bury me in the Falls"). In 1891, *Century* magazine published a story on the climb of Mount St. Elias, in which the writer describes a risky moment: "At one point we found a snow-bank jutting forward over the bench we wished to reach, and although it was a sort of Sam Patch game we made it, cutting steps up the incline to the ice on the other side." Among the other publications in which Patch's name appeared were *Scribner's, Atlantic Monthly, Putnam's, Living Age,* and even *Manufacturer and Builder.* Forty years after his death a book entitled *The Wonderful Leaps of Sam Patch* was published and included yet another long poem of praise.[12]

It seems incredible today that Patch's leaps acquired such a mythic quality. Had no one jumped from a high cliff into water before? Certainly others must have. The fact that Patch got people to pay for his first jump, and that he was able to parlay this stunt into a full-scale career complete with vast media coverage, says less about the uniqueness of his act than it does about the fixations of the public at the time. The Jersey Jumper caught fire simply because the world was ready for him. As with the parachutist Garnerin and his imitators, and the parade of gravity performers who had filled the circus in the previous century, something about the times caused this kind of falling to have a strange mass appeal. If Patch were the only notable falling daredevil of the age, this might not be true—but, as we'll see, he was only the first of dozens who achieved international celebrity in the nineteenth century.

A similar case was that of Samuel Gilbert Scott, also a diver, who took up where Patch left off, in the 1830s and 1840s. Scott was an American sailor from Philadelphia who began by diving from the topmasts of ships. He also made a jump from Niagara Falls, which was reported in a newspaper as an impossible 593-foot dive (Niagara Falls itself is less than half that height; besides, a fall of that magnitude would have certainly killed him). In 1840, Scott went to England, where he would post broadsides describing himself as "the celebrated American diver" who would "precipitate himself from an eminence of upwards of 140 feet."[13]

Scott's leaps, like Patch's, became legendary and were probably exaggerated. His record, if it can be believed, remains unmatched today. In Brighton, he dove from a 140-foot scaffold; in Liverpool, it was 167 feet; in Devonport, from the 200-foot mast of the battleship *St. Joseph*; in Cornwall, from a 240-foot cliff. (This last dive is equivalent to jumping from San Francisco's Golden Gate Bridge, a fall that has killed more than a thousand people, and Scott reportedly did it into eight feet of water. It's highly unlikely he made this jump; today, the world's record for a high dive officially stands at 177 feet. It was set in 1984 by Oliver Favre-Bulle, who broke his back when he hit the water.)

Like Patch, Scott perished during one of his stunts, although in an even more dramatic fashion. To enhance his act, Scott would tease the crowd before a jump by grabbing a rope and swinging by his feet or neck. On January 11, 1841, he was doing just that when the noose slipped and tightened around his throat. His audience mistook this for a trick and applauded as he hung to death. Naturally, this bizarre end only added to Scott's myth, and shortly after his death one newspaper was so desperate for a picture of him that they published a handy portrait of the poet Robert Burns and put Scott's name under it.

High divers like Patch and Scott not only revealed the public's growing appetite for gravity stunts but also pushed the image of the gravity hero further into the realm of rugged individualism. Unlike the generations of circus performers in Europe, these Americans weren't tied to any show-business traditions whatsoever. They presented themselves simply as courageous, raw men ready to meet gravity on its own terms.

And it was this last element that was perhaps the key to their appeal. The high divers' impact on the Victorian imagination was greater than that of circus acrobats because they confronted the terrible threat of nature more directly. While wirewalkers and leapers did show a mastery over nature-as-gravity, they did so in controlled conditions. Moreover, the point of their performances was to *avoid* falling, which meant they never actually plunged into gravity's lair. They taunted the dragon but didn't slay it. Patch strode out to make jumps through the windy air, over unknown rocks and into churning rivers, actually allowing himself to be *swallowed* by nature, returning almost literally from the belly of the beast. To the Victorian mind, this was an almost biblical demonstration of individual prowess.

In case anyone might miss the point, Patch had made it plain that he should be regarded as a conqueror worthy of great admiration. "Napoleon was a great man and a great general," Patch said before his final jump at Genesee Falls. "He conquered armies and he conquered nations, but he couldn't jump the Genesee Falls. Wellington was a great man and a great soldier. He conquered armies and he conquered nations and he conquered Napoleon, but he couldn't jump two Genesee Falls. That was left for me to do, and I can do it well."[14] Patch's motto was "There's no mistake in Sam Patch"—no flaw in his being that would stop his willed triumph over gravity.

These claims to heroism, though, didn't wash with the monitors of morality. Immediately after his last leap, moralizers let loose a tirade. The *Rochester Gem and Ladies Amulet* made the message plain: "Sam Patch, who had rashly, but till now uninjured, sported with the law of nature, has given us an example that vain and mortal man may not trifle with bounds prescribed by an Omniscient God."[15] Pastors railed from the pulpits, their sermons reflected in this editorial from the *Rochester Observer:*

> We have all, more or less, been accessory to this awful daring of the providence of God. . . . We are always in His keeping, but are never to

> dare His power. This exploit would have been extolled in shouts by the gazing multitude, had he risen in safety from the abyss; but an incensed God had decided otherwise. . . . All who gave countenance to them, by witnessing or encouraging their exhibition, offend against the majesty of Heaven, and are daring the vengeance of an offended Deity.[16]

The fire-and-brimstone rhetoric found its target, and many of those who had gladly put in a coin or two to watch Patch were suddenly filled with deep shame. "In our imagination," one witness recalled, "we often see him hanging in mid-air, casting a reproachful look at those who by their presence encouraged his feats of frenzy."[17] Another local newspaper described the feeling:

> It was a daring and useless exposure of human life, which, having resulted disastrously, creates a train of painful reflections. We would not dwell upon this distressing scene, and yet we cannot banish it from our thoughts. We still see the frail mortal, standing, as it proved, upon the brink of eternity! The terrified imagination follows him from the giddy height, through the thin air, into the deep, dark chasm below! It lingers but a few moments of breathless and agonized suspense! The waters, troubled a moment in swallowing up their victim, are at rest! The expiring bubbles announce that the spirit has departed, leaving the body in the "dark bosom of the ocean buried." The multitude shrinks away abashed and rebuked.[18]

In spite of their guilt—or perhaps because of it—many didn't want to believe that Patch had actually perished. The fact that his body wasn't found immediately offered hope. He was sighted for months after his last jump, in bars and on village streets throughout New England. One man turned up in Albany claiming he *was* Patch, but this was quickly disproved.

The spring following Patch's last leap, however, put an end to the mystery. When a ranch hand broke an ice sheet to water his horses, Patch's body floated to the surface, still in his white uniform with a black scarf around his waist.

Neither Patch's nor Scott's death stopped the parade of high divers. In 1852, *Scientific American* reported a jump at Niagara Falls—"a second Sam Patch leap," as they described it (declining to name the jumper)—adding that "such feats appear to take their course like fashions. One fellow has been amusing us New Yorkers for three weeks past, with jumping off the High Bridge into the Harlem River—It is a profitless and dangerous feat. Sam Patch lost his life at last, and Scott, the celebrated American leaper turned crazy and put an end to his life on London Bridge."[19]

And yet the gravity heroes kept coming.

## Unjustifiable Curiosity, Brutal Apathy

Back in Europe, parachutists were already upstaging high divers as popular gravity heroes—and like them, they would be admired by the masses and denounced by self-appointed moral authorities.

The veneer of parachute drops as scientific demonstrations had quickly worn off, and this had to do with the nature of falling. For the public, the sheer spectacle of a man falling safely from two thousand feet instantly overshadowed the picture of the parachute as an advance in aeronautic science. Although Garnerin seemed to fancy himself an amateur scientist and referred to at least one of his later parachute displays as an "experiment"[20] (he repeated his performance in Paris once a year from 1798 to 1800, and then one final time in England in 1802), he was, in reality, a showman. He made his living as a performing balloonist, traveling as far as Russia to play for paying crowds; parachuting was just an occasional part of the show. He even got his fiancé, Jeanne Labrosse, into the act. In 1799, she became the first woman parachutist, fluttering down from three thousand feet after launching from Tivoli Gardens in Paris.

Moreover, the parachutists who followed Garnerin didn't bother with any scientific pretensions at all. They developed no new designs but simply copied Garnerin's, took their shows on the road, and charged admission. By 1819, balloonists-turned-parachutists had performed in Russia, Austria, Germany, Portugal, and America. André Garnerin's own brother, Jean-Baptiste Garnerin, began to mount his own profitable shows and realizing that a woman falling added sex appeal, didn't drop himself but instead had his daughter Elisa jump. From 1815 to 1836, she made thirty-nine leaps in venues as far off as Prussia, Spain, and Italy, where she awed a crowd of sixty thousand. Appalled at his brother's crass commercialism, André refused to speak to him. But André's opinion, even as the first in the field, did nothing to stop the parachute's swift turn toward entertainment.

Experts in the newly emerging science of aeronautics were repulsed by the parachutists' exploitation of public voyeurism. Their sentiments were echoed in an editorial from the London *Sun*, commenting on Garnerin's jump in 1802:

> This is the first experiment of the kind in this Country, and we sincerely hope it will be the last. We mean not to detract from the skill and the courage displayed by M. Garnerin upon this occasion; indeed, it would be impossible if we were so inclined; but the man who could feel any pleasure in seeing the life of his fellow creature exposed to such imminent danger, *without any adequate cause* [italics added], must possess either the most unjustifiable curiosity, or the most brutal apathy.[21]

To right-thinking minds, parachuting looked less like a scientific demonstration and more like a new, barbaric form of entertainment.

So science all but ignored the new invention, in spite of the fact that a practical use for it was obvious from the beginning and deserved research. Because hot-air balloons were lifted with fires in their baskets, they had the nasty habit of bursting into flames in midair, leaving their pilots with no safe way down. (Jean-François Pilâtre de Rozier, who had been one of the first two men to fly in a balloon in 1783, unwisely yoked a hydrogen and hot-air balloon together when he tried to cross the English Channel two years later. The craft exploded and killed him.)

Using a parachute to escape, it seemed, could have saved a lot of lives. Yet in its entire first century, it was never developed for that purpose. This would have required some sort of collapsible canopy that a balloonist could keep on board as a safety device, but the early parachutists were reluctant to risk any changes to Garnerin's model (even though it oscillated horribly—Garnerin actually vomited from motion sickness after his "experiment" in England). And they were reluctant for good reason. The first test using a new design meant to eliminate the sickening oscillation—a design based on the best aeronautic science of the time—had ended in disaster.

ROBERT COCKING WAS a professional painter, but like Garnerin, he considered himself a man of science at heart. After seeing Garnerin in 1802, this Briton had become fascinated with parachutes and later had experimented with miniatures based on an inverted cone design suggested in 1810 by Sir George Cayley (who would later gain fame by proposing the idea of fixed-wing aircraft). He had even given lectures on the design to the City Philosophical Society and the Society of Arts. The inverted design used a canopy that looked liked Garnerin's rigid, parasol-shaped parachute turned upside down: its tip pointed downward. The idea was that the air, instead of being trapped under the canopy (which is what caused the oscillation), would instead flow up the slope of the canopy, while still offering enough air resistance to slow the fall. It was a sound idea, as Cocking had proved with his miniatures.

Cocking envisioned a full-size model capable of carrying a man, but a balloon large enough to lift what would be a huge, heavy parachute had never been built. Then, in 1836, it was built. The enormous, eighty-foot-tall aircraft had been commissioned by the owners of Vauxhall Gardens, and had been dubbed the Nassau Balloon when it was flown 480 miles from London to the Duchy of Nassau. After twenty-six years of waiting, here, finally, was the balloon that could lift the design Cocking had in mind.

Cocking, the amateur scientist, may have had qualms about getting involved with the common showmen who owned the balloon. But at sixty-one, the aging painter may have also thought it was his last chance to realize his dream. So a deal was struck. If the owners would cover the expenses, Cocking would make the jump without pay. He set to work on his parachute, building the frame with tin tubing to keep the enormous structure as light as possible.

The broadside advertising the show was a mixture of scientific boast and carnival barking. Like the *Titanic*, advertised as "unsinkable," Cocking's new design was proclaimed utterly failure-proof:

> The Proprietors of Vauxhall have the satisfaction to announce that they are enabled to present to the Public another grand improvement connected with the Science of Aerostation; viz. A PARACHUTE of an entirely Novel Construction, by which a perfectly safe and easy descent may be made from any height in the Atmosphere attainable by a Balloon.
>
> MR. COCKING, a gentleman of great scientific acquirements, having, many years since, witnessed the descent of M. Garnerin, (the only one ever made in England,) was forcibly struck with the danger to which that gentleman was exposed on account of some error in the construction of his machine; and, after several years spent in numerous experiments, has succeeded in discovering the faults in M. Garnerin's instrument, and also in producing
>
> AN ENTIRELY NEW PARACHUTE,
>
> which is allowed by all who have seen it, to be constructed on unerring principles. The form is that of
>
> An Inverted Cone 107 Feet in Circumference!

which, during the Descent, is quite free from oscillation; and as it will be in its proper form previous to the Ascent, it is not liable to the objection of falling several hundred feet without expanding, which was the case with the Parachute of the old form.

In order to render this Fète more than usually attractive, the Proprietors intend giving a variety of Amusements during the Afternoon, the principal of which are—

A CONCERT in the Open Orchestra
A DRAMATIC PIECE in the Theatre, which will be lighted as at Night
The Extraordinary Performance of M. Latour, M. Dela Vigne, & their Sons
THE YEOMANRY AND QUADRILLE BANDS. &c. &c. &c.
AND A VARIETY OF OTHER ENTERTAINMENTS.[22]

On a warm July afternoon in 1837, Londoners streamed to the banks of the Thames and into the Vauxhall Gardens' arcades and groves, milling along its grand concourse among its arches and pavilions, dining on sweetmeats and wine, waiting for the big event.

At 7:30 P.M., the Yeomanry Band launched into the national anthem. The great Nassau balloon lifted off, with Cocking's parachute—a huge, shallow cone painted with garish decorations—dangling beneath it.

At five thousand feet, Charles Green, the balloon's pilot, shouted down to Cocking that they had better release the parachute, otherwise it would be too dark for the crowd to see the descent. He asked how Cocking felt. "I never felt more comfortable or more delighted in my life," Cocking replied. Green, not as confident in the experiment as Cocking, asked the parachutist if he was sure he didn't want to be lifted back up into the balloon's basket. "Now, Mr. Cocking," he said, "if your mind at all misgives you about your parachute, I have provided a tackle up here, which I can lower down to you, and then wind you up into the car by my little grappel-iron windlass, and nobody need be the wiser." The implication was that they would send the parachute down unmanned. "Certainly not!" Cocking replied.[23] After saying their good-byes, Cocking pulled the cord that freed his craft.

Nothing happened. Cocking then wrapped the cord a few times around his wrist, to get a better grip. He yanked again. The latch released, and the balloon shot up. Cocking, with his arm still wrapped by the release rope, was violently jerked up until he hit the parachute's frame. This snapped the release rope. He fell back into the basket, but the frame had now been bent. In a moment, as the air gathered under his rigid canopy, he slowed, and all seemed well in spite of the accident.

But then above him, the tin struts began to shudder, as wind from below grew to a blast. Something creaked. The struts then buckled; the canopy lurched to one side, then to the other, and finally was crushed. Blown upward, it was now flapping like a shredded jib in a gale. Cocking's world began to spin as the crumpled canopy twisted, tightening the lines to his basket and sending him plummeting in a tight spiral. The wind came roaring like a hurricane.

With nearly all air resistance gone, Cocking would have reached terminal velocity in under fifteen seconds, falling at a hundred miles an hour or more. He would still have another three thousand feet—about another fifteen seconds—to understand his fate before he hit. Witnesses said that within the last three hundred feet, the wind ripped his mangled canopy from the basket, sending it naked to the ground.

The first men to reach Cocking pulled him from the wreckage and carried him, still breathing, to the nearby Tiger's Head inn. He died in minutes. As a crowd gathered, the inn's owner charged threepence to see the wreckage, sixpence to see the old man's broken body.

THE ACCIDENT WAS blared across the front pages of the tabloids the next day; some even printed special "Parachute Disaster" supplements. For the next several years, opinion pages railed against parachutists. The *Examiner* of London, commenting on a successful demonstration made by John Hampton in 1839, compared parachuting to the most brutal entertainment of the time:

> Now what is there in prize-fighting, or bull-baiting, more really savage than this, or betraying so much callous inhumanity in the spectators? Here was an immense throng of people whose pleasure consisted in seeing a fellow creature risking his life with every probability of losing it, in a way the most horrible to the imagination.[24]

Like Garnerin's success, Cocking's failure was played out before a crowd, and so it had an equally intense but inverse effect on the public. The thrilling image of a successful parachute descent—the first time people had seen a man fall that far and live—was now fused with a picture of intense horror: the first time people had seen a man fall that far and *die.*

## The Flying Trapeze

In the early 1800s, with Patch and Scott taking hundred-foot dives and parachutists dangling thousands of feet overhead (and in Cocking's case, demonstrating the danger by perishing), classic gravity acts—those of the circus—had begun to look contrived, perhaps even safe. But around midcentury, two traditionalists, Jules Léotard, an acrobat, and Jean-François Gravelet, a wirewalker, set out to steal back the gravity hero spotlight.

Léotard's fame came suddenly and seemingly out of nowhere. Born in 1838, he had been raised in Toulouse in southern France, near the Pyrenees. Since his family's business was a public gymnasium (where his father taught physical education), Léotard had grown up with the gym as his playground. One day, as he was about to plunge into the indoor pool, he glanced up at the row of ventilators in the ceiling. Each ventilator had two long cords, about a foot apart, dangling from it—cords used to open and close the vents. And each set of cords was separated from the next by several feet. The dangling cords reminded him of those that support a pair of rings or a trapeze, both of which were fairly recent additions to standard gymnasium equipment. But seeing several pairs of cords in a row caused him to imagine an innovation: several trapezes hung in a row, lined up as the ventilator cords were. If this were done, he wondered, would it be possible to swing from one trapeze to another?

Since his father owned the place, young Léotard rigged up just such a set of trapezes to find out. And he arranged them over the pool so he could fall without injury. After much practice, he found that he could not only swing and grab the next trapeze but also actually *fly* some distance between them. By 1859, the twenty-one-year-old was happily accomplished on his invention, unaware that he had created what would become the definitive aerial circus act.[25]

A few circus performers who visited the gym saw the potential of Léotard's "flying trapeze" immediately, and they were in a position to do something about it. They invited him to Paris, to the Cirque Napoléon, to perform. On November 12, 1859, Léotard appeared at the great Parisian circus, and because there had been no hint that such an act had been developed, he stunned the audience and became a phenomenon.

His act kept getting better. At the Cremorne Pleasure Gardens in London, in 1861, he hung five trapezes in a row—the most ever. The band struck up a waltz, and Léotard appeared in a bright, one-piece body stocking. He grasped the first trapeze and swung, and then instead of just flying from one trapeze to another, he turned somersaults between them, directly over his astonished audience. He became so well known that his costume was dubbed the "leotard."

One reason Léotard's act enthralled people was his technique, which was different from that of today's trapeze artists, who let go at the top of their swing to be caught by someone who's also swinging. Witnesses said he would let go near the bottom of his swing and literally fly to the next swinging trapeze. You can see why he had to: letting go at the top of his arc would have dropped him nearly straight down, unable to catch the next bar and with no forward momentum. To hit the next bar with enough force to keep swinging—and enough space to somersault—Léotard had to release before he rose too high. Using that technique, he flew across gaps of a dozen feet or more.



Trapeze acts caught fire with the Victorians, bringing them in droves to see Richard Beri and James Leach, acrobats from the St. Petersburg circus who had converted to be trapeze artists, and Mlle. Azella, a woman who also performed in a leotard—considered scandalous for a woman—and who dismounted her platform with a double somersault from thirty feet into a mattress. Léotard himself died of smallpox ten years after he invented his stunt, but the trapeze flourished. By 1868 flyers were performing double flips from trapeze to trapeze, and in 1870, a swinging catcher became a regular addition.

Spectators were spellbound not just by the performers' skill and courage but also by the raw sensuality of it all, sensuality that was enhanced when heroic men joined heroic women to challenge gravity together. The young bodies in skin-tight costumes, working up a sweat, with limbs akimbo, slamming into each other as brawny men caught slender girls, raised the Victorians' pulse in more ways than one. One performer, described by Arthur Munby, was an Englishwoman of eighteen. "The only clothing she had on was a blue satin doublet fitting close to her body and having very scanty trunk hose below it," he noted. "Her arms were all bare; her legs, cased in fleshings, were as good as bare, up to the hip."[26] As her partner hung by his knees from a trapeze, she slithered down over his chest, let herself drop, then caught her feet in his armpits. She crawled back up by wrapping her legs around his body.

For the finale, she was to swing across a gap of eighty feet, through two paper discs, and into the arms of her male catcher; as Munby wrote, "A fair girl of eighteen, preparing in sight of all men for such a feat as that; perched up there, naked and unprotected, with no one to help her; anxiously testing the ropes, chalking the soles of her feet, wiping the sweat off her hands and her bonny face, and trying to smile withal. . . . She did it, of course . . . and was seen hanging in the arms of her mate, grasping his body, her face against his breast."[27]

The sexism of the description, although expected for the time, reveals another aspect of the evolving image of the gravity hero. Had that account been written by a woman instead of a man, it's likely that the "girl" wouldn't have been seen as "unprotected, with no one to help her," but instead as the protagonist in the drama—after all, *she* was the one flying through the air. We might even speculate that although they dare not voice it, women in the audience might have been seeing in female trapeze artists, to use today's term, an *empowered* woman—a gravity hero of their own.

## The Great Blondin

Léotard's invention had suddenly upstaged wirewalkers, and there seemed no way they could come up with anything as thrilling as somersaulting through the air. But there was. The key would be to add greater risk. The man who realized this, and who became one of the most famous entertainers of the nineteenth century because of it, was Jean-François Gravelet, better known as the Great Blondin. Whether he was inspired by Patch's and Scott's direct challenge of nature isn't known, but nevertheless he understood that bursting the confines of the circus tent would not only add the needed danger but also make room for a lot more paying customers. And so he traveled to America, and to the site of Patch's greatest (and most profitable) performance: Niagara Falls.

Blondin's first performance was on June 30, 1859. Just downstream from Niagara Falls, grandstands had been set up on the edge of the cliffs overlooking the river. The roar from the Falls was constant even at that distance, and the mist made the hot air muggy for the hundreds of people who sat jammed on the hard wooden bleachers. Beyond them, a crowd lined the cliffs for a half mile up and down the river, twenty-five thousand people standing as close as they dared to the edge to get the best view.

Blondin, as some in the crowd knew, had begun walking tightropes as a five-year-old and had turned professional by the time he was nine. The little Frenchman was already well known in Europe. Now, at thirty-five, he had plenty of experience. But as the crowd looked at the rope, which stretched more than a thousand feet across the canyon, its diameter about that of a coffee cup, they must have wondered if even the most skilled wirewalker could make the trip without falling. The rope wasn't even level: it dipped far down in the middle, so that Blondin would have to walk on a downward slant and then uphill to finish. The rope would be held steady—that was clear from the dozens of guy wires that ran at angles to the rocks below—but that wouldn't stop the wind. If he was blown off, the 160-foot fall would kill him. If by some miracle it didn't, the rapids would. Even though Blondin was said to be a strong swimmer—he had once jumped from the deck of a ship to save a passenger—he couldn't hope to survive Niagara's raging whitewater.

Blondin's physical appearance didn't make his chances look any better. A small, delicate man with blond hair and goatee, he seemed almost feminine in his purple vest, white pantaloons, and flesh-colored tights, as he lifted his huge balancing pole and stepped out onto the rope. But once he was on the rope, it was clear that he was master of it. He walked casually, as if it was routine. In less than five minutes he had reached the sag in the middle. And there, he stopped.

Those who were watching through opera glasses could see that he was fooling with something now: a bottle. He tied a cord to it and lowered it to a boat heaving in the river, the *Maid of the Mists*, where a deckhand took it. The sailor dipped the bottle into the river, filling it with water, and then let Blondin pull it back up. The wirewalker took a swig and smiled. Regripping his balance bar, he began to walk up the slanted rope, just as easily as he had before. On the far side he could be seen waving. Someone handed him a glass of champagne, which he drank before stepping back out on the rope. This time he did a little jig before strolling back to his starting point, completing the round-trip in under eight minutes and without once appearing to be in the slightest danger.

A few weeks later, unable to believe what they had read in the papers about a man crossing above the Niagara on a tightrope, an even larger crowd gathered to see his performance: forty thousand people packing themselves along the canyon's edges. This time when he stepped out onto the rope, he was not holding a balancing bar—and the crowd soon discovered why. After walking for a bit, he paused, stood with his arms outstretched, bent his knees, and hurled himself into a back flip, landing perfectly, one foot in front of the other. The crowd erupted in applause—but Blondin was just getting warmed up.

He next appeared on the rope carrying a chair. After walking three hundred feet out, he set the cross braces of the chair on the rope, sat down, crossed his legs, and serenely surveyed his dumbfounded audience. Blondin then stood, picked up the chair, and continued across the rope.

A little farther along, he again took the chair from his shoulder. This time he put just two legs of it on the rope, letting the other two hang in space, and again took a seat as if in his parlor, seemingly bemused at the horrified crowd. Now he was up again and casually strolled along before stopping for a third time. Again he put only two chair legs on the rope. But this time he stepped onto the chair and stood—on a chair balanced on two legs on a rope 160 feet above the raging river.

At that point, witnesses report that women began to faint.

THROUGHOUT THE SUMMERS of 1859 and 1860, while Léotard thrilled Europeans with his new gravity trick, Blondin continued to astonish Americans at Niagara with his extreme version of wirewalking. His tricks became ever more brazen and remain scarcely imaginable today. He crossed at night, crossed blindfolded in a sack, crossed with his legs and arms shackled, crossed on fork-shaped stilts, did handstands, and hung from the rope by an arm or leg. He invited his manager to climb onto his back and carried him all the way across, barely managing to shove through the waiting crowd, grown so thick that they threatened to knock the pair into the gorge.

Near the end of his first season, Blondin strapped a small cast-iron stove to his back, along with a skillet, eggs, a whisk, and a spatula. He balanced the stove on the rope, lit a fire, broke the eggs into the skillet, whipped them around, and created an omelet, which he then lowered to the *Maid of the Mists* below. People on board the boat were frantic for scraps of the omelet, some shoving them into their mouths, others carefully wrapping pieces as souvenirs.

IN A PATTERN that was becoming typical with gravity performers, the reaction to Blondin and Léotard was that admirers heaped on absurd praise while critics unsheathed their knives. The Lockport *Chronicle* said Blondin's walk was "the most terribly real and daringly wonderful feat that was ever performed." No less a source than the *New York Times* called it "the greatest feat of the nineteenth century."[28] On the other hand, *Punch*, in 1862, had this to say:

> The taste for seeing fellow creatures put their lives and limbs in danger we cannot call romantic, but view rather as disgusting. It is not so much the skill of the performer that attracts audiences, as the peril he is placed in and the chance of seeing his neck broken. If monkeys could be trained to do the tightrope and trapeze business, they would soon eclipse the feats of Blondin and Léotard. Monkeys are by nature better fit for such achievements and having fewer brains than men, have no fear of falling. Surely, we repeat, it would be a good thing for humanity if acrobatic monkeys could be trained and exhibited. The lives of human beings then need not be endangered, and the public might be weaned from its present brutal taste for seeing men imperil their existence by attempting feats which monkeys could achieve with perfect safety, and far more ease and skill.[29]

It was ridiculous, though, to expect that the public could be "weaned" from these demonstrations by watching animal stunts or anything else. The crowds came because what gravity performers accomplished seemed similar to other wonders of human progress: men sailing ships across seas like fish, floating balloons skyward like birds, and rolling locomotives across the plains like horses. But it was also different, and perhaps even more impressive, because it involved little, if any, new technology. Blondin and Léotard's mastery of gravity was driven entirely by will. It was the soaring spirit of the individual that made proponents cheer and detractors curse: the crux of the matter was that here were men armed with nothing more than determination and ability triumphing over one of nature's primary forces, the great pull of the earth itself.

Mountaineering exploded in popularity in the mid-nineteenth century, when virtually all the major western European peaks were climbed with almost no mishaps. This Golden Age ended with a shock in 1865 when mountaineering's greatest triumph of the day, the conquest of the Matterhorn, also brought its first horrific disaster, the fatal plummet of four of the seven climbers. This depiction was drawn by Gustave Doré, circa 1865. *(Getty Images)*

CHAPTER 3

# Trouble in the Theaters of Gravity

Another Niagara Crank Disposed Of. —*New York Times*, on the death of rapids barrel-rider Robert Flack, 1885

SAM PATCH MAY have opened Niagara Falls as a venue for gravity performers in 1829, but Blondin's performances in 1859 suddenly made it a main stage. One reason is that while there were few professional high divers to follow Patch and Scott (and fewer still after they vividly demonstrated the risks), there existed scores of wirewalkers, and after Blondin's success, it seemed that every one of them wanted to stretch a rope over the Niagara River.

The first who dared to compete directly with Blondin during his second season in 1860 was Signor Guillermo Antonio Farini, or "The Great Farini" as he was known—or William Hunt as his parents had christened him, for Farini was actually Canadian. But with his tan face, goatee, and long black hair, his looks certainly fit his Italian stage name. Far from having a show-business background, as had Blondin, Hunt's strict parents had forbidden him from even going near the circuses that passed through his hometown of Port Hope, Ontario—although he would sneak in anyway. At twenty-two, Farini had taught himself everything he knew about rope walking by copying performers he had seen, practicing in his own backyard.

Which only made his accomplishments more impressive. After bragging that he would match or outdo Blondin, amazingly he did. Blondin balanced on a chair? Farini balanced on his head. Blondin hung from an arm? Farini hung from his toes. Blondin carried out a stove and cooked an omelet? Farini carried out a hundred-pound washtub and did laundry. Blondin carried his manager across the cable? Farini carried a larger, heavier man—and carried him not across a taut cable but across an unsteady slack rope. Farini's legs trembled so during the stunt that the crowd shouted he had no right to risk the man's life.

And in a gravity feat that stands as one of the most difficult of the era, or indeed of any era, Farini completely upstaged his rival. Even Farini didn't realize what he was getting into until it was almost too late.

On August 15, 1860, he set out along a slack rope stretched over the canyon. On his shoulder he carried a coil of rope. When he reached the center, he fastened his balancing bar to the slack line on which he stood, and dropped one end of the rope he carried to the *Maid of the Mists,* where it was caught and held by a deckhand. Farini tied his end to the slack line. Then Farini gradually climbed down the rope, all the way to the *Maid's* deck.

Once onboard, he drank a glass of wine. He then gripped the rope with both hands, wrapped his legs around it, and began to drag himself back up. Farini was incredibly strong—he could lift seven hundred pounds and once won a tug of war against a dozen men—but this was a *two-hundred-foot* rope climb, through wind, on a rope tugged by a lurching boat and affixed to an unsteady slack line above.

At only a third of the way up, his hand and arm muscles were nearly tapped out. He then could climb only ten feet at a time before stopping to rest, twisting his legs around the rope and shaking out his arms. By the time he reached the top, his arms were dead. He forced his chest over the slack rope and just hung there.

Finally, he put a leg over the rope and sat on it. He unfastened his balancing bar but couldn't lift it—couldn't even really hold it. Using his toes, he coaxed the bar along his shins to his knees, then raised his knees so the bar was near his chin. He managed to grip it with one hand—his other hand was on the slack rope, desperately trying to keep it from swaying to and fro. He got one foot on the line and then the other, and trembling with weakness, slowly stood up. Yet when at last he reached safety at the canyon's edge, his ordeal was not over. Farini had advertised that he would make the return trip blindfolded and with baskets on his feet. After only a ten-minute rest, that's just what he did. The crowd that had come, as always, for the chance to see a wirewalker fall had nearly seen it—although they had no idea that they had. No other wirewalker—and especially not Farini—attempted the rope-climbing trick again.



BOTH BLONDIN AND Farini made money, but Farini, who turned out to be an extremely astute businessman, made a *lot*. Instead of relying on contributions, as had Blondin, Farini had hired schooners and trains to bring people in and had arranged for long bleachers to be placed up and down the cliffs, so he was able to sell tickets for forty thousand seats. From one performance he made more than $15,000—about $325,000 in today's dollars.

Blondin and Farini and the wirewalkers who followed all understood the terrifying magnetism of Niagara that made audiences willing to shell out good money for their shows. As a correspondent for the *Times* of London put it, "one half of the crowds that go to see Blondin go in the firm expectation that as he must fall off and be lost some day or other, they may have the good fortune to be there when he does."[1] It was a simple formula: the greater the danger, the bigger the crowds. The danger could be increased in two ways: by trying stunts that made falling more likely, and by making the fall itself seem more horrible—that is, making the human seem more vulnerable or nature more powerful. Either way, the drama was heightened, and the audience grew.

Blondin and Farini used both approaches. Blondin steadily increased the difficulty of his tricks throughout 1859. For his second season in

1860, he augmented the horror of his possible fate by moving his wire farther downstream; this put him just upriver from the violent rapids that would sweep him toward certain death in Niagara's quarter-mile-wide whirlpool if he fell.

Farini, too, worked his way up to more difficult stunts in his 1860 shows. And after both he and Blondin left the Falls, Farini returned in 1864 with a trick that had nothing to do with wirewalking but everything to do with increasing risk: using iron stilts, he announced, he would wade out in the waters *above* the Falls to the very lip of the cataract.

On the day of the stunt, Farini stepped into the river from the Goat Island Bridge, walking in the shallows toward American Falls. Halfway to the precipice, the river broke one of his stilts, twisted his ankle, and threw him into the rapids just a few hundred yards from the drop. Farini struggled onto one of the small islands just above the Falls, where he was rescued a few hours later.

Farini then left Niagara for good, but he went on to become an extremely successful gravity impresario. He quickly jumped on the trapeze bandwagon that Léotard had begun, and in 1866 opened in London with El Niño, his eight-year-old son who could fly forty feet from one trapeze to another. When that show had run its course, he presented another trapeze act: Lulu, who was actually El Niño disguised as a woman and who thrilled audiences with a triple somersault from trapeze to net.

In 1877 Farini invented the human cannonball. His performer, a fourteen-year-old girl named Zazel, was launched from a spring-loaded device of his design, through a barrel, timed to coincide with an explosion. Zazel flew about thirty feet into a net. Imitators followed, each trying to fly higher and farther, and by the mid-twentieth century human cannonballs were using compressed air to shoot themselves two hundred feet in an arc that could reach a height of a hundred feet. The force typically caused them to lose consciousness for a second or two in flight, but they usually regained it in time to maneuver into a landing position. The act requires almost no skill but is extremely risky: of the people who performed it into the 1950s, two-thirds were killed when they missed the net. On the last of her performances, Zazel hit the ground, and the impact broke her back. She remained crippled for the rest of her life. Farini again moved on and as we will see in a later chapter, became instrumental in the history of the parachute.

Blondin, after his seasons at Niagara, toured the world, appearing before thousands, often with a painted backdrop of the Falls behind him. He performed until he was seventy-two and retired a wealthy man. At home, he still enjoyed an occasional stroll across the rope stretched in the backyard of his mansion in Ealing, England, at the corner of Niagara and Blondin Avenues.

The wirewalkers who followed Blondin and Farini to Niagara tried to wring what they could from the danger-equals-crowds formula. In 1865, Harry Leslie strung his rope even farther downriver than had Blondin, so he walked directly *over* the rapids. Henry Bellini, in 1873, crossed the gorge on the longest tightrope ever used—fifteen hundred feet—offering that much more opportunity for him to slip and fall. In 1893, Clifford Calverly impressed crowds by actually running across his tightwire, completing the trip in two minutes and thirty-two seconds.

There were others who filled the years between Blondin and the end of the century, many of whom performed for several seasons: Professor J. F. Jenkins, Stephen Peer, Maria Spelterina (the only woman), Samuel Dixon, James Hardy, Charles Cromwell, and D. H. McDonald. But the thrill wirewalkers could muster began to fade long before they stopped coming. It faded because so many had proved by now that crossing the canyon just wasn't that hard. Only Stephen Peer had trouble; in 1887, he fell and died—but this happened not in front of a crowd but at night and probably while he was drunk. His body was found below his rope, on the rocks near the river. And one of the wirewalkers—Harry Leslie—had actually *jumped* from his wire into the river three times during his performances in 1873, which seemed to negate the risk of walking over the river in the first place.

By the late 1800s, people at the Falls had tired of wirewalkers. There just wasn't enough gravity in their gravity stunts. And as anyone could see, Niagara Falls was all about the awesome power of gravity.

IF THERE'S ONE place on earth to witness the magnitude of gravity's power, it's Niagara Falls. Only Victoria Falls, in a remote section of Africa, is wider. Niagara is the single runoff point for four of the great lakes, which together contain one-fifth of the world's fresh water; it is the place where tourists can see more mass falling than anywhere else.

Niagara's might had long fascinated people. As early as 1728, artists had tried to capture the Falls in sketches and paintings, and when these seemed too small for its monumental spectacle, they created dioramas and huge, painted panoramas: one canvas covered five thousand square feet. A stage play used ten thousand gallons of water to try to replicate the Falls. In 1853, Godfrey Frankenstein produced a panorama of Niagara that moved: a hundred paintings on three gigantic rolls of canvas. Each roll would be slowly unfurled across the stage of his Broadway theater, each painting, in turn, coming to rest behind a huge frame, as live music played and Frankenstein himself described the scene. Here was a view of the calm river above the Falls; here were the rapids, trembling just before the cataract; here, the furious mountains of water plunging over the Falls themselves, so real you felt swallowed by the froth and spray. Looking at this terrible display of gravity's power, who didn't imagine, at least for a fearful moment, what it would be like to be carried away by that maelstrom—to actually go over the Falls? Who wasn't transfixed by stories of those who had?

Tourists who visited the Falls heard the tales from guides who delighted in pointing out where the unfortunate had been swept to their doom. There was the one about Francis Abbot, the hermit of Goat Island, who wandered near the edge of Horseshoe Falls barefoot, sometimes with a violin in hand, and who one day in 1831 was seen to neatly remove his clothes, fold them, and slip into the water above the Falls. His body was found below them eleven days later. There was the case of Nettie De Forest, a little girl on vacation with her parents and family friend Charles Addington. On a hot July day in 1848, Nettie slipped off the rocks of Luna Island, at the crest of Bridal Veil Falls, and though Addington grabbed her, he fell in as well and they both went over, clutching each other as Nettie's parents watched in horror. In 1844, Martha Rigg fell into the boiling waters from a ledge. Almost immediately, an entrepreneur set up a stand to sell souvenirs and poems about her death. He was so successful that a competitor moved in, and the two men fought over placing their stands at the exact spot of her fall.

The latest of these stories began to unfold the same year Frankenstein's painted panorama did. In July of 1853, Joseph Avery, whose boat had overturned upriver, was sighted clinging to a log trapped in the rocks near American Falls. His two companions had already been swept over. In the city, readers opened their *New York Times* to follow the nineteen-hour rescue attempt in successive editions. First rescuers floated a raft to Avery; this took several tries, but eventually he managed to get on it. When the raft became trapped in the rocks, a lifeboat was sent down to him. But the lifeboat knocked Avery off the raft, and the crowd on the bank watched as he was washed to the edge of the Falls, where he thrust up a hand, screamed, and went over. Frankenstein added a new, tragic painting to his show.

In light of these morbidly fascinating stories, anyone could see that what remained for daredevils to do was to somehow challenge the Falls themselves, as had last been attempted by Farini with his stilts. In 1889, there arrived at Niagara a man who had an idea of how that could be done.

## Crossing Niagara

Arthur Midleigh was a wealthy Englishman who had come to America for adventure. He had originally wanted to fight Indians but discovered that at the end of the century there were few hostile tribes left. Accomplishing something at the Falls was his last hope for excitement. In talking with the rivermen, he soon learned that no one had yet dared row across the river above the Falls, and in this he saw his chance to make his mark.

Arthur Midleigh, though, didn't realize that the Niagara River is deceptive. For miles above the Falls, the river flows wide over land that is dead flat. It begins to tremble into rapids only a few hundred yards upstream from the Falls. Near its northern shore, where Midleigh planned to launch and where he stood to judge its current, the rapids run relatively slowly and shallowly; from this view he simply couldn't see the much faster flow in the center.

What Niagara hid from him was her enormous liquid muscle, greater than that of the Columbia River, submerged in the deep channel that runs through her heart. Here, nearly a hundred million gallons a minute surge through with such force that it makes a sieve of the rock walls that contain it. All around the cataract the stone is shot through with finger-size holes, holes the spurting water pries open into cracks that eventually send acre-sized chunks of land tumbling from the cliffs. Across the entire two-thousand-foot breadth of the Falls, the river chews off three feet of rock every year; at the Falls's base, it has gouged out a pit two hundred feet deep. As Midleigh stood on the shore watching the river's undulating skin and hearing only its gentle flow against the rocks, he declared that he had rowed through waters that ran twice as fast.

Midleigh, though, knew he would never be able to manage the trip alone. Over the next few days he scouted for a partner, offering fifty English pounds—half the price of a modest house in those days—to any riverman who would go with him. In spite of the generous fee he had no takers. Eventually Midleigh turned to Alonzo Gardner, the young man who had been acting as his guide around town, and by upping his offer to a hundred pounds he got Gardner to sign on.

On a September morning, a crowd gathered on both sides of the river to see the latest challenger to Niagara. A few hundred yards upriver from Goat Island on the American shoreline, Midleigh and Gardner stepped into a rowboat, gripped the oars, and began cutting through the chop toward the center of the channel.

Almost immediately the river's swells took hold of the boat and began to toy with it, hurling it along at forty miles an hour, the bouncing skiff sending spray cascading over the men. Suddenly realizing the power of the river, they desperately tried to row their way out, digging their oars into the river's shoulders to skew toward shore. But Niagara bore them down, driving them along the shore of Goat Island into the torrent that led to the American Falls. A few hundred feet from the edge, the river slammed the craft into a rock. Midleigh and Gardner dragged themselves to safety while the waters upended their boat and sent it over the drop.

As darkness fell, the two men clung to the rock while rescuers tried to decide what to do. The plan they hatched was to find a boat, tie it to a line, and let the current take it to the stranded men. But this had to wait until daylight, and so the pair was left to spend the night without sleep, drenched by the river's spray, while the roar of the Falls beckoned from only a few yards away.

It took all the next day to find a suitable boat, and by the time it was brought to the shore, dusk was coming. Once again, Gardner and Alonzo were forced to spend a miserable night on their tiny island.

As the third day dawned, the two could see several thousand people who had now gathered on the shore. The rescuers shoved the boat into the river and let out the line. Niagara carried it along her rapids to the stranded pair; just then, the rescuers on shore clenched the line and dug their heels into the bank to hold the boat against the river's grip. As if to torture them, Niagara lifted the vessel and smashed it against the rocks, splintering it into a useless carcass.

Next the rescuers floated down a large wooden beam and held tight as it tossed and heaved among the whitecaps, crashing into the rock.

Midleigh stood, steadied himself, and then jumped for the beam. He caught it, but the river rolled the beam out from under him and then pulled him beneath the boiling water and spat him out over the Falls. At that moment, the crowd saw Gardner fall to his knees and pray. It was dark again.

Gardner spent his third night on the rock. At dawn's light he looked across the water to see twenty thousand people lining the shore. Again, the rescuers floated a beam to him, this time with a harness that he was to put on and then hook to the head of the beam. But the river carried the beam swiftly past the exhausted man. Fifty volunteers took the rope in hand and strained to drag the beam back within Gardner's reach. In a final act of contempt, Niagara took the beam and tossed it into the air, smacking Gardner off his feet and into the cauldron. She then washed him to the brink. Witnesses said Gardner lifted his torso up out of the river and clutched at the sky just before the river took him over.

MIDLEIGH AND GARDNER had challenged the Falls all right, but instead of becoming gravity heroes they had simply become victims in another lurid tale in which Niagara and gravity had won. But the idea of beating the Falls remained. Late in the century, there was talk in the taverns among rivermen about the possibility of going over the cataract in a barrel. They talked of a barrel for a simple reason: in 1886, a cooper named Carlisle Graham had come to town and rode a specially constructed barrel he had made through the vicious rapids below the Falls—and survived. In fact, he had done it four times over the years. And just before Midleigh and Gardner met their fate, Graham had claimed he had taken his barrel *over* the Falls—although this was exposed as a lie.

Graham, though, *had* sent the barrel over, and it had withstood the shock—so the idea seemed plausible. Who would be the first to try it? Maybe Graham would do it, to clear his name. Perhaps it would be Bobby Leach, a Niagara stunter originally from Cornwall, England, who had also ridden through the rapids in a barrel.

But no one was willing in 1889, the year of Midleigh and Gardner. Niagara would wait a dozen more years before the arrival of its signature stunt, and as it would turn out, neither Graham nor Leach nor any of the other regular Niagara rivermen would be first. The first instead would be someone no one could have predicted.

## "I Done It Once"

Although Niagara Falls was a mecca for gravity daredevils, it wasn't the only one. Even before the Brooklyn Bridge was completed, it was targeted by leapers who thought they would become rich by being the first to plummet from its railing. The bridge stood at just about the perfect height for jumpers; at 130 feet, it was a formidable risk, but not high enough to make death or injury certain. And as an international landmark, it guaranteed headlines.

The first to try was Ronald Donaldson in 1882. He was stopped once by wind and twice more by bridge construction workers. After his attempt, authorities became watchful, and they were on guard when, three years later, Robert Odlum made a public announcement that on May 19, 1885, he would jump.

Odlum wasn't a professional daredevil; he was a swimming instructor from Washington, D.C., and hoped to make enough money to build himself a swimming pool. Knowing the police would try to stop him, he arranged for a diversion. As thousands gathered to watch, he had a friend go to one end of the bridge and pretend to jump, which distracted the cops long enough to give Odlum his chance. He arrived on the bridge's opposite end dressed in a bright red swimsuit. He pointed one hand straight up, locked the other to his side, and stepped off. In the air, he tilted. He hit the water flat on his back and was instantly killed.

Odlum may have been wrong about his chances of survival, but he wasn't about his chances for wealth. A year after his jump, Steven Brodie said he would succeed where Odlum had failed. On July 23, 1886, a group of witnesses announced that Brodie had indeed survived a leap from the Brooklyn Bridge. Brodie became famous overnight.

We can get a glimpse of his celebrity through the eyes of Nikola Tesla, the great physicist, electrical engineer, and inventor, who describes in his autobiography what happened one day when he, evidently joking, claimed he had jumped from the Brooklyn Bridge:

> Steve Brodie had just jumped off the Brooklyn Bridge. The feat has been vulgarized since by imitators, but the first report electrified New York. I was very impressionable then and frequently spoke of the daring printer. On a hot afternoon I felt the necessity of refreshing myself and stepped into one of the popular thirty thousand institutions of this great city, where a delicious twelve per cent beverage was served, which can now be had only by making a trip to the poor and devastated countries of Europe. The attendance was large and not over-distinguished and a matter was discussed which gave me an admirable opening for the careless remark, "This is what I said when I jumped off the bridge." No sooner had I uttered these words, than I felt like the companion of Timothens, in the poem of Schiller. In an instant there was pandemonium and a dozen voices cried, "It is Brodie!" I threw a quarter on the counter and bolted for the door, but the crowd was at my heels with yells, "Stop, Steve!", which must have been misunderstood, for many persons tried to hold me up as I ran frantically for my haven of refuge. By darting around corners I fortunately managed, through the medium of a fire escape, to reach the laboratory, where I threw off my coat, camouflaged myself as a hardworking blacksmith and started the forge. But these precautions proved unnecessary, as I had eluded my pursuers. For many years afterward, at night, when imagination turns into specters the trifling troubles of the day, I often thought, as I tossed on the bed, what my fate would have been, had the mob caught me and found out that I was not Steve Brodie![2]

Immediately after the jump, Brodie earned a $100 by displaying himself as the "world's most courageous man" at a Brooklyn museum. He opened a tavern in the Bowery with a painting over the bar that showed him plummeting; nearby was a framed statement from the barge captain who claimed he had fished Brodie out. The tavern was a huge success and attracted celebrities. Before the 1892 boxing match between John L. Sullivan and James J. (Gentleman Jim) Corbett, Brodie let his customers know he thought Sullivan would knock out Corbett in the sixth round. Corbett was a patron of the bar, and one day brought his father in to meet the famous Mr. Brodie. Feeling protective of his son, the elder Corbett was in no mood to be impressed.

"So you're the man who jumped over Brooklyn Bridge," he said.

"No, no," Brodie replied. "I didn't jump over it; I jumped off it."

"Oh," snorted Corbett's father. "I thought you jumped over it. Any damned fool could jump off it."[3]

By 1889, Brodie was such a well-known daredevil that like Sam Patch before him, he next went to Niagara Falls. In early September of that year, just before Arthur Midleigh attempted his fatal crossing, Brodie heard the news that Carlisle Graham had become the first human to go over the Falls in a barrel. A few days later, Brodie announced that he had duplicated Graham's stunt. Brodie's fame grew and was such that in 1894 he starred in a Broadway musical, *On the Bowery*, in which he sang and reenacted his plunge from the Brooklyn Bridge (with the help of some special effects), this time to rescue the play's heroine. *On the Bowery* was an enormous hit; it toured the country and made Brodie rich.

Brodie's success, though, was based on a fraud. According to most scholars, he hadn't jumped from the bridge at all but probably had a friend throw a dummy while he slipped into the water from below to be "rescued." The only other record of Brodie jumping from a bridge is a report in the Poughkeepsie *Daily Eagle*, dated 1888, in which he's described as having jumped from the Poughkeepsie train trestle, after which he was rescued by friends and taken to New York to recover. If so, it would have been a long recovery: the bridge is 212 feet high—a fall nearly impossible to survive. Asked many times over the years why he wouldn't repeat his stunt, if only to quell the controversy, he always replied, "I done it once." Brodie was also caught in a lie—he had not gone over Niagara Falls in a barrel.

Despite Brodie's Niagara fraud, his fakery of the Brooklyn Bridge jump couldn't quite be proved, and he was able to live off his reputation until he died in 1901, at age thirty-six, of diabetes. Like Patch, Brodie's name lived on. It became synonymous with leaping; later, stuntmen would refer to high falls as "brodies." A 1940s action actor, John Stevenson, would take "Steve Brodie" as his stage name. Brodie's story was depicted in a 1933 film, *The Bowery*, which starred George Raft as Brodie. Brodie even made an appearance as a character in a Bugs Bunny cartoon, in which Bugs tricks him into jumping.

The first man to actually jump from the Brooklyn Bridge and survive, a month after Brodie claimed to have made his jump, was Larry Donovan, a press operator who went on to jump from the upper suspension bridge over Niagara Falls. Donovan was apparently trying to parlay his fame into a career, but his plans were cut short when the following year he leaped from London Bridge and was killed.

In 1887, at least nine people jumped from the Brooklyn Bridge; after that, the leaps declined but never disappeared completely. What the jumps foreshadowed wasn't an increase in high-diving daredevils, because these jumpers weren't professionals. Instead, they would portend the coming of a more disturbing phenomenon: that famous high structures, like the Brooklyn Bridge and later the Golden Gate Bridge, would soon become a gravity theater of a different kind: stages on which thousands would play out their final act.

## Alpine Fever

During the same years that Blondin, Farini, and the early wirewalkers were invading Niagara, mountaineers in Europe were rapidly turning the Alps into a theater for high-risk climbing—and falling. And like Niagara, the phenomenon was driven by artists' images, tourism, and sensational journalism.

For the first decades after Balmat and Paccard's ascent of Mont Blanc in 1786, mountain climbing in the Alps remained for the most part something only scientists were inclined to do. But as the Industrial Revolution put more money into the pockets of a new leisure class, Alpine resorts, particularly at Chamonix and Zermatt, were soon overflowing with tourists. At the same time, fine artists, inspired by the romantics' exaltation of nature, began to imply that the simple act of getting up a precipice had new meaning. J. R. Cozens, J. M. W. Turner, and Eugène von Guérard all made mountains central to their pictures. In *Wanderer Above the Sea Fog*, an 1818 painting by Caspar David Friedrich, we see a climber from the back, a large figure in a long dark suit coat, standing in the dead center of the canvas. He looks to have just arrived on this jagged peak, one foot perched on the edge, the other just behind, his cane braced between his hip and a rock. We look out, as he does, to crags shrouded in fog, islands jutting up through the clouds. It is an image of a monarch surveying his conquered realm.

That realm was seen as spiritual. It looked as heaven might, and if God was above, these places were certainly nearer to him. John Ruskin wrote in his influential book *Modern Painters* that he hoped depictions of mountains could be used to show a link between heaven and earth. Actually, they had often represented just that in the background of religious paintings since the Middle Ages. Now, though, they were foreground, with mortals climbing heavenward. Like a mirror image of the impulse that had caused moralizers to see Sam Patch's plummets as an affront to God, mountaineers would eventually see climbing almost as an act of religious devotion.

But the reverence of painters, and of romantic writers such as William Wordsworth (who had praised the Alpine scenery), merely primed the pump of recreational mountaineering. The spigot was opened by the authors of travel guides, who in successive editions would describe routes that according to one contemporary review, "formerly were never thought of by any mere tourist, [but] are now methodically described, so that it is possible to anticipate to a great extent the time, the fatigue, the comparative danger, and the expense of almost every ascent which ever has been made, at least in the more frequented parts of Switzerland."[4] The flow was increased by the city fathers of Chamonix, who in 1823 organized the local guides into a guild and established fixed rates, which made choosing a peak to conquer as easy as selecting an entrée from a menu. And finally, the floodgates were opened by the hyperbole of a showman, the urgings of an attorney, a priest, and a bon vivant, and the seduction of a scientist.

The showman was Albert Smith. A pudgy English writer, Smith had been obsessed with the Alps since he had been a boy. In 1851, after climbing Mont Blanc with an enormous entourage (they took ninety-one bottles of wine, four legs of mutton, and forty-six fowl, among other necessities), he published a breathless account, and the next year presented "The Ascent of Mont Blanc," an elaborately illustrated lecture.

Like Frankenstein's Niagara panorama, Smith arranged his illustrations on rolls that slowly unfurled; at dramatic points, he would have his brother stand behind them with a candle to light up the moon. Smith put on a full show; he sang, waxed poetic over every view, and used every gimmick he could think of: live Saint Bernard dogs, Alpine milkmaids in costume, and a skittish chamois imported for the occasion. He also exaggerated every danger. "The speaker," as one review put it, "abandoned his jokes and puns, and became terribly serious as he described the horrors of the final climb, that being a matter much too serious for even a professional wit to touch without, as the reports say, being 'visibly affected.' "[5] His show became a phenomenon; Smith performed before packed houses in London for eight years and played three times for the royal family. He stirred such interest that a board game based on climbing Mont Blanc was produced, ensuring that every feature on the mountain was familiar to children throughout Britain.

The attorney was Alfred Wills; the priest and bon vivant were, respectively, the Reverend Charles Hudson and Edward Shirley Kennedy. Wills was wealthy and respected—in later years, as a judge, he presided at the trial of Oscar Wilde—but in 1854 he was in Switzerland, gripped by the idea of conquering a new peak. Hudson (vicar of Skillington in Lincolnshire) along with Kennedy (a full-time adventurer, thanks to a family fortune) set their sights on Mont Blanc. The books they published—*Wanderings among the High Alps*, by Wills, and *Where There's a Will, There's a Way: An Ascent of Mont Blanc by a New Route and Without Guides*, by Hudson and Kennedy—proposed a radical new idea: that mountain climbing wasn't a form of scientific expedition but a sport.

And a hell of a sport it was. The Alp Wills went after was no gentle dome; it was the Wetterhorn, a massive block of a mountain topped with a three-peaked crown of summits. Near the top, Wills clung to a seventy-degree snow slope as his guide hacked through the face of an enormous frozen wave, the wind-blown cornice at the peak. The last chunk fell, and suddenly there was sky. They scrambled through the opening and to the summit:

> The whole world seemed to lie at my feet. The next moment I was almost appalled by the awfulness of our situation. The side we had come up was steep; but it was a gentle slope, compared with that which fell away from where I stood. A few yards of glittering ice at our feet, and then, nothing between us and the green slopes of Grindelwald, nine thousand feet beneath.[6]

Hudson and Kennedy's climb of Mont Blanc was hardly as thrilling, but they offered readers a different temptation. By climbing without guides, they proved you didn't have to be wealthy to be a mountaineer—and that plenty of climbs, including Mont Blanc, were actually *easy*: "the risk of serious accident was but little greater than incurred by a pedestrian in the streets of London,"[7] they claimed.

Hudson and Kennedy in particular, as a contemporary reviewer put it, "had shown that the ambition of getting up hills, the excitement of encountering danger in the Alps, and the interest of skillfully surmounting difficulties, were a sufficient inducement in themselves"[8] for the new sport.

The scientist was John Tyndall. Serious physicist though he was (he published on subjects ranging from magnetism to the properties of light and sound), this rail-thin Irishman fell completely under the spell of mountaineering. Fully half his book titled *The Glaciers of the Alps*, published in 1860, was devoted to vivid descriptions of his climbs and of the intoxicating mix of elation and fear climbing produced. Here's Tyndall on the summit of Monte Rosa:

> I thought of my position: it was the first time that a man had stood alone upon that wild peak, and were the imagination let loose amid the surrounding agencies, and permitted to dwell upon the perils which separated the climber from his kind, I dare say curious feelings might have been engendered. But I was prompt to quell all thoughts which might lessen my strength, or interfere with the calm application of it.[9]

The books were hugely popular, and suddenly the rush for "peak-bagging" was on. In 1857 Kennedy and some friends formed the Alpine Club, which ballooned from eighty members to nearly three hundred in three years. And while nearly half were certifiably upper crust—attorneys, clergymen, landed gentry, university dons, and the like—more than half weren't. Mountaineering, indeed, was for everyone.

Wills's conquest of the Wetterhorn (which, although he didn't know it, had been climbed at least a few times before) was later seen as the beginning of what would be called the Golden Age of Mountaineering. In one two-year span, nearly a hundred first ascents would be made; within the eleven years of the Golden Age, virtually every high Alpine peak would be conquered. On nearly all of these climbs, there would be no accidents. But it was only a matter of time before a fall of horrifying proportions would occur. When it did, its circumstances could hardly have been more dramatic.

## The Most Infamous Fall

The final prize of the Golden Age was the Matterhorn, the site of the tragedy. The Matterhorn, surely the most recognizable mountain in the world, is a gigantic, sharp, four-sided tooth sticking fourteen thousand feet up into the sky, familiar to anyone who has seen its replica at a Disney theme park. Set apart from surrounding peaks, its rock sides are so steep that snow looks like a brushstroke of thin white paint. To the locals, the fact that it couldn't be climbed was obvious just by looking at it.

The stunning profile of the Matterhorn, and the look of the Alps in general, soon supercharged the synergy between popular publications and climbing that had been sparked by Wills, Hudson, Kennedy, and Tyndall. In the early part of the century, illustrations had appeared in newspapers and magazines for the first time, and by the 1840s, metal engraving had made the pictures vivid—and cheap to produce. In 1860, when Blondin was making headlines with his walks across Niagara's gorge, London publisher William Longman, looking to exploit the new Alpine fever, sent an artist to the Alps, a promising young engraver named Edward Whymper.

Whymper embodied what the Swiss had detested in the English for decades. He was arrogant, bullying of guides, and disgusted by the "squalid" villages—all this at only twenty years old. He hated Switzerland, even the scenery, but in short order he discovered that he loved climbing and that he was good at it. He attempted the Weishorn but didn't make it; the next year he came back and made the first ascent of Mont Pelvoux. And then he aimed for the great prize: the Matterhorn.

He went up with one guide and naturally got in a fight with him when the guide refused to climb a dangerous chimney; they retreated. Whymper was furious. His arrogance, as would be revealed in his next season, wasn't without justification: as a climber, he would soon be ranked among the best in the world.

In 1862, Whymper tried the Matterhorn twice with guides and failed both times. Frustrated, he then attacked the mountain alone, and astounding other mountaineers, including veteran John Tyndall (who was also aiming to be first on the peak), he climbed higher than anyone had been before. Not only that, during his descent he proved just how tough a mountaineer he was.

On the way down, he slipped, fell twelve feet into an ice chute, slid down its length until he was launched sixty feet through the air, hit the chute's rocky side, slid again, and stopped himself just ten feet short of an eight-hundred-foot drop. Cut and bruised everywhere, he plastered a handful of snow to the four-inch gash in his head, climbed up to a crevice, and passed out cold. He awoke at dusk and limping and bleeding, climbed back up the entire chute and then four thousand feet down the mountain—at night.

It was an astonishing piece of work merely to have survived such a climb, and it served notice that Whymper had more skill and stamina than dozens of mountaineers who had been climbing for years. Yet he was just getting started. After only a few days' rest, he went up again but was forced to retreat because of the timidity of his guide, who was scared off by a snow flurry. So he went up again, with a new guide, and climbed even higher than his own record-setting point, "but at length we were both spread-eagled on the all but perpendicular, unable to ascend, and barely able to descend."[10] He returned to Breuil, got a ladder to bridge the cliff that had stopped him, along with a new guide, and went up *again*—his sixth assault on the Matterhorn. Once more, he failed to find a route that would work.

When Whymper returned to the village, he discovered that Tyndall was now on the mountain. Alone, he climbed up once more, this time just to see what his rival was up to. He climbed so fast he actually passed Tyndall—but then came down: he now knew better than to attempt summiting by himself. In the meantime Tyndall almost reached the top but was thwarted by a hundred-foot notch along the final ridge. Dejected, he gave up completely. "The Matterhorn is inaccessible," Tyndall announced, "and may raise its head defiant as it has hitherto done—the only unconquered and unconquerable peak in the Alps."[11]

It didn't matter to Whymper that Tyndall was among the most respected mountaineers of the age; he publicly ridiculed the veteran for not knowing the notch was there and for not planning for it. Tyndall might believe the Matterhorn couldn't be climbed, but that was just because he hadn't tried hard enough. Whymper all but said that Tyndall had simply chickened out.

The following year, 1863, Whymper made another attempt but was beaten back by a twelve-hour blizzard in which he and his guide were nearly hit by lightning. He didn't try in 1864—the weather on the mountain was bad—but instead decided to knock off a few more "impossible" climbs. He scaled the shoulder of the Meije, thought unclimbable by the locals. He made the first ascent of Pointe des Écrins and descended by a route that had him straddling a thin ridge threatening a four-thousand-foot drop at one point and leaping eight feet from a rocking boulder across a deep chasm at another point. And on he went: first ascent of Mont Dolent, of Aiguille de Trélatête, and of Aiguille d'Argentière, and also across unconquered passes and up perilous ice chutes, every one a vicious, dangerous climb.

All the while he was sending a constant stream of sketches back to London, sketches that were a revelation for the public. Now added to the vivid descriptions of Tyndall, Kennedy, and others was the nineteenth-century version of raw photo journalism. Whymper's images were action shots, brimming with shock value. In one of his engravings, a companion is caught in midair, falling from a steep snow ledge to a snow bank thirty feet below, flying face down, while his three stunned partners look on. In another, Whymper himself is the flyer; he's fallen backward from a knife-edge ridge and is shown plummeting down the cliff. Everything in

Whymper's drawing heightens the viewer's dread: his arms are spread wide, rocks fall next to him, his hat and alpenstock float yards above him, a streak of shadow against the wall throws him into relief, and if there is a fluffy snow bank awaiting him, Whymper has chosen to exclude it from the bottom of the frame.

Back again in 1865, Whymper tried a new route up the Matterhorn. It didn't work. It was yet another failure from the south side, the "back" of the Matterhorn, hidden in most pictures we see of it. He had always climbed from the south, as had nearly everyone else, because it sloped far more gently than the north side, which rose like a wall above Zermatt. Looking up to its massive triangular face, he now envisioned yet another route: climbing along its northeast ridge. But he would have to act fast: a group of Italian climbers had already mounted another assault from the south, along the Italian ridge.

Pushed by competition, Whymper didn't have time to put together his own group, so he instead joined another Englishman in Zermatt who had the same idea. Lord Francis Douglas, only eighteen, had done a little climbing, but he was hardly a mountaineer. He had already hired two guides, Peter Taugwalder and his son (also named Peter). By coincidence, the Reverend Charles Hudson, coauthor of *Where There's A Will, There's a Way* and by now an expert mountaineer, was also planning to climb the Matterhorn and was staying at the same hotel as Whymper. Hudson had with him Douglas Hadow, a nineteen-year-old with no climbing experience whatsoever, yet the good reverend had promised to take him, along with their own guide, Michel Croz. Whymper suggested all seven of them go together, a suggestion that would be their undoing.

On July 13, 1865, they set off. The first day took them on a moderate hike up to eleven thousand feet, where they set up camp for the night at the base of the ridge. The guides spent the afternoon scouting the route, and to their astonishment the way looked far easier than they had expected.

They started at dawn the next day. For three thousand feet the climb was like walking stairs; most of the time they didn't even bother to rope themselves together. Even so, Hadow constantly needed help. By midday they reached the high shoulder of the mountain, where its tip angled up sharply and where the ridge was no longer climbable. They stepped out onto the silent face of the mountain, a forty-degree rock slope that below became a nearly sheer four-thousand-foot wall. Here snow caked the slope, and the few nubbins of exposed rock that offered holds were filmed with ice—"Something altogether different,"[12] Croz announced. They roped up with Croz leading. Clinging to the rock with hands and the toes of their hob-nailed boots, they worked their way up. That stretch, though, was short and gave way to a gentle snow bank leading to the top. Croz and Whymper took off running and reached the summit together at 1:40 in the afternoon. When they looked down, they saw their rivals, the Italians, far below.

The peak of the Matterhorn turned out to be the size of a football field, and for an hour they congratulated themselves, soaked up the sun, looked out over Switzerland to the north, where blue smoke rose from chalet chimneys, and to the south, over the sunlit pastures of Italy. The day was clear and bright. Whymper was still making sketches and chatting with Taugwalder's son when the four others of the group began planning for the descent.

Croz, the strongest climber, was again set to lead. After him came Hadow, the weakest. Next was Hudson, then came young Lord Douglas, and finally, the elder Taugwalder. They all thought this would be the safest order, the guides on either end and Hudson in the middle providing anchors in case one of the two amateurs roped between them fell.

Whymper and Taugwalder's son lingered as Whymper wrote their names on a note and stuffed it in a bottle to mark their achievement. By the time the two caught up with the others, Croz and his ropemates were already gingerly moving down the slick face that led to the easier climbing on the ridge. Whymper roped himself to Taugwalder's son, and the pair began down the slope, unroped to the leading group.

Earlier Whymper had suggested that they anchor the rope to a rock for the descent, but his idea was ignored. Among the leading five, only one man would move at a time, and they evidently thought this was precaution enough. A little farther down, Whymper roped himself to Taugwalder at the end of the leading group, so that now all seven were connected.

At a comparatively easy part of the route, Croz turned to make sure Hadow, just behind him, was secure, actually taking the teenager's feet in his hands and placing them on the rock. Croz then turned to go down—and Hadow slipped. His feet skidded out from under him and punched Croz in the back. Croz cried out, toppled over headfirst, and sailed into the air; Hadow slid after him. The rope snapped taut. It easily yanked Hudson and Douglas right off the face.

Adrenaline shot through the last three men. Whymper and Taugwalder's son, the last two in line, froze in their stances, muscles rigid, praying that Taugwalder could stop the chain reaction. As Douglas flew away from him, Taugwalder clamped his arms around a rock in terror, pressing his face full into it. The rope slithered in the air, straightened, snapped tight, thinned, and finally broke. Taugwalder's grip had held. The old guide felt a hard jerk, then nothing. The rope lay slack.

The view down was clear. Whymper watched the four lost men sliding on their backs, spreading their arms, legs, and hands, launching off slabs of rock, then dropping silently against ice crags that shredded their bodies, all the way down the four-thousand-foot face.

"For more than two hours afterward," Whymper wrote later of their descent, "I thought almost every moment that the next would be my last." Their nerves were shattered, courage to move even a few steps was gone, though they now tied their ropes to the rocks. "Even with [the ropes'] assurance," Whymper wrote, "the men were afraid to proceed, and several times old Peter turned with ashy face and faltering limbs, and said, with terrible emphasis, '*I cannot!*' "[13]

AN ONSLAUGHT OF publicity, fueled by the fact that disaster had come so soon after success, followed in the aftermath of the climb. A young boy in Zermatt had actually seen the fall through a telescope, and the villagers knew of it the instant the survivors staggered in. Rumors rocketed from the tourist-filled village; one said that Tyndall had been killed. Whymper's account was published in the London *Times* on August 8, and then the world knew, creating the nineteenth-century equivalent of a media frenzy. With it came ethical questions about guides leading unskilled climbers, and accusations of who did what when and why.

Whymper reported that the rope that had broken was the weakest they had, a backup, one that should never have been used if it could be avoided, as it was no thicker than a clothesline. People accused Peter Taugwalder of selecting it on purpose, hoping it would break if those in front of him fell; many even said he had cut the rope. Whymper knew this wasn't true—he had seen the rope's frayed end—but he was unsure about why Taugwalder would have picked the old rope. A court inquiry was held, but it was inconclusive, cheating Taugwalder of vindication and leaving him shunned by his neighbors for the rest of his life.

For the public, the combination of events surrounding the fall—that it was on the Matterhorn, the most dramatic looking and last of the Alps to be climbed; that it came at the end of a decade of mountaineering accomplishments; that among the party was Reverend Hudson, practically a celebrity author; that the fall itself came after summiting; and crucially, that a journalist capable of both illustrating and telling the tale was there—made it a tragedy of mythic proportions.

It was depicted over and over, by Whymper and many others, for decades. Forty years after the disaster, Ferdinand Hodler, known for monumental paintings of historical events, painted two murals, *The Ascent* and *The Plunge*, showing in exquisite detail the men struggling upward and then plummeting, eyes open, gasping, wreathed in swirls of rock and snow as they drop. Looking at the paintings, it's hard not to sense the familiar theme of hubris against heaven, of pride coming before a fall.

The fall stretched to hideous new dimensions what people understood a fatal fall to be. Mountaineers had fallen, but no one had fallen like this before, not four thousand feet over meat-grinding ice and rock. The Matterhorn had shredded and ripped the clothes off the men; their bodies were found not only mangled but also humiliatingly, nearly nude—a horrific sight to prudish Victorians. Even toughman Whymper was appalled, writing later, "I have never seen anything like it before or since, and do not wish to see such a sight again."[14] The men had been slashed and pummeled almost beyond recognition, like victims of a frenzied, vengeful killer: the top of Croz's head had been sheered off; he was identified by his rosary cross, which the mountain had jammed into his jaw so hard it had to be pried out with a knife.

The Golden Age of Mountaineering abruptly turned to lead and sank. Suddenly, mountaineering didn't seem like a noble adventure so much as a madman's—perhaps a sinner's—compulsion. The press furiously condemned mountaineering. "Is it duty? Is it common sense? Is it allowable? Is it not wrong?"[15] railed the London *Times*. Ruskin declared to mountaineers, "You have made racecourses of the cathedrals of the earth. The Alps, which your own poets used to love so reverently, you look upon as soaped poles in bear gardens, which you set yourselves to climb and slide down with shrieks of delight."[16]

And there it was, in Ruskin's words, the ancient screed against gravity challengers, complete with the implication that it was unholy, and a reference to bear baiting (that's what went on in "bear gardens") that recalled early-eighteenth-century critics' insistence that that's where vulgar tumblers and rope dancers (and now mountaineers) belonged.

Perhaps those who found the bodies, including Whymper, were given pause by the same feelings when they buried the climbers at the foot of the mountain. As they laid them to rest, reading Psalm 90 from Hudson's prayer book, which was found near his body, they may have winced at these lines:

> *For we are consumed by thine anger, and by thy wrath are we troubled.*
> *Thou hast set our iniquities before thee, our secret sins in the light of thy countenance.*

Construction innovations developed during the nineteenth and early twentieth century filled cities with towering structures for the first time, but also brought an unforeseen consequence: thousands of people determined to vault over the railings into oblivion. Workers on the Golden Gate Bridge, shown here under construction in 1937, could hardly have guessed that their gorgeous span would shortly become the most popular place in the world to die. *(Photo by Ted Huggins. California Historical Society; FN-09260)*

CHAPTER 4

# Sirens of Height

He would hear them screaming as they jumped. He believed they changed their minds halfway down.

—WIFE OF A PAINTER ON THE GOLDEN GATE BRIDGE

DURING THE EIGHTEENTH and nineteenth centuries, mountaineering, parachuting, roller coasters, and daredevils changed the way people felt about falling by displaying both its glory and its horror. Yet there was more. It was a gradual change but farther ranging: a trend in architecture that would expose more people to higher places than ever before. Structures were becoming taller, and tall structures more common.

The trend's ultimate expression was envisioned in 1884, by Gustave Eiffel and engineers in his company, as the centerpiece of the World's Fair that was to take place in Paris in 1889. Their aim was to build the tallest structure on earth. The Eiffel Tower, when completed, would be 986 feet tall, almost twice the size of anything that had come before it. It would be made of 18,000 individually forged pieces of iron and would weigh 76,000 tons.

Eiffel worked through every engineering problem with meticulous precision, but halfway through construction he was stopped dead by a very human reaction to such height: his workers became terrified of falling. It didn't make any sense to Eiffel. A fall from his tower was no different from one off any of the high places these men had worked before. "The professional risks remained the same," he wrote, "whether a man fell from forty meters or three hundred meters, the result was the same—certain death."[1] Although that was true, Eiffel was overlooking a psychological effect that occurs whenever a human imagines a fall: we sense how long it will take. Our "terrified imagination" plummets more than seven times longer from three hundred meters than from forty, multiplying sevenfold the terror we would expect to feel in the air. As with mountaineering, a fall of this magnitude was new.

No workers in history had been asked to work at heights like this, because quite simply nothing this size had been built before. The Great Pyramid of Cheops, built around 2580 B.C. and measuring 481 feet, stood as the tallest structure on earth for nearly four thousand years. In 1307, the central spire of the Lincoln Cathedral in England topped Cheops at 525 feet. Over the next five centuries, a few isolated churches and cathedrals occasionally peaked at over five hundred feet, but they were exceedingly rare.

What kept a lid on tall structures before the modern era were two things. The first was that everything known about building was based on trial and error. Any ambitious architect going beyond what had been tested was simply guessing, and chances were good that the structure might suddenly collapse, as some of the taller cathedrals did a century or two after they had been erected. The second was that timber, stone, and brick were only strong enough to build something so high before it imploded under its own weight, no matter how carefully it was built. Using iron would have solved the problem, but that was out of the question: smelting even a small quantity took huge amounts of charcoal and hours of labor.

All this changed during the eighteenth century. In 1709, Abraham Darby substituted coke for charcoal, and in 1760 John Smeaton invented the blast furnace—suddenly iron was cheap. Builders also had new mathematical techniques based on the work of Newton that let them calculate the stress on a structure. By 1779, the first bridge made entirely of cast iron was built, and the era of tall public structures began. When E. G. Otis installed his first elevator in 1857, he guaranteed a future of multistoried office buildings filled daily with people working hundreds of feet from the ground. Public heights soon became places from which anyone might fall by accident. In essence, cheap iron would create a new exposure to falling, and along with it, a newly heightened fear, one first voiced by the protesting workers who vexed Eiffel.

These men were spending up to twelve hours a day pounding hot rivets into place with heavy iron hammers, balancing on girders scarcely wider than their shoes, with six hundred feet of air between them and the ground. They insisted that the higher they climbed, the more frightened they felt—and they wanted more money for it. But Eiffel wouldn't have it. "For a true construction worker," he said, "vertigo does not exist."[2]

Eiffel broke their resistance by offering a bonus of a hundred francs to every man who would work until the French flag was raised on the tower's pinnacle, and by firing all who instead held out for an hourly raise. He hired new workers and sent them right up.

For those who had agitated for the raise in the first place but now accepted his new offer, Eiffel reserved a clever punishment: he *restricted* them to the lower level, saying he would fire them if they disobeyed. In effect, he branded them with their fear of falling. These men were—literally—looked down on by their coworkers, and most quit.

And so Eiffel's tower rose, with no more outbreaks of acrophobia. "Thus," he said, "it was proved that with the proper equipment a good construction worker can work at any height without feeling unwell."[3]

ACROSS THE ATLANTIC, engineer Augustus Roebling was making his own contribution to civilization's abundance of high structures by revolutionizing the way bridges were made. Just as the innovations that made iron cheap made possible huge structures like Eiffel's tower, Roebling's startling new methods would create bridges higher and longer than any the world had seen.

Roebling accomplished this by inventing a way to manufacture—on site, if necessary—the huge steel cables that would make possible enormous suspension bridges. When he began, respected engineers ridiculed his ideas because they didn't trust suspension bridges of any size in the first place. In Europe, five built with chains had collapsed, and so had one built with cable, sending a brigade to a canyon's floor. In America, cables the size Roebling was talking about were unknown.

Roebling first proved his methods by building suspended aqueducts, held up with cable he manufactured himself. These worked so well that he was commissioned to provide cables for Charles Ellet, who was building a railroad bridge over what nearly every other engineer had said was an impossible chasm—that of Niagara Falls. To designers of the time, the idea that a train could be held up with wire seemed like a recipe for disaster.

After Ellet was fired in a financial scandal, Roebling was asked to design the entire bridge—and here, he presented a radical idea. Suspension bridges up to that time had been designed to be flexible, to absorb, it was thought, the stress of the wind. Engineers didn't realize that there was another danger: vibrations, either from wind or even from marching soldiers, could resonate in an unreinforced suspension bridge until they were amplified to a power that would tear the bridge apart. In 1854, this happened to one of Ellet's bridges, shortly after a witness had stepped from it. (Incredibly, engineers made the same mistake nearly a century later in building the Tacoma Narrows Bridge across Puget Sound. Its 1940 collapse was captured on film, the roadway bucking into huge waves under a light wind that set off just the right vibration, causing the bridge to writhe until it burst.)

Roebling worked with fanatical attention to detail, failing even to return home for the birth of his fourth child. After its design was proved by withstanding a twelve-hour gale during construction, the bridge opened in March of 1855. That day a train weighing 368 tons rolled across. (It was from Roebling's bridge that spectators would later watch Blondin and many of Niagara's wirewalkers cross the canyon.)

With this success and others behind him, Roebling was the obvious choice for another impossible project, what would be the largest, longest clear span in the world: the Brooklyn Bridge. It, too, would be a two-tiered bridge, this time of enormous proportions: the span would be half-again longer than his own record-setting Cincinnati Bridge.

Roebling supervised every detail, including the construction of a massive machine that would braid the cable on the spot. In 1869, when construction was about to start, Roebling's foot was crushed when a ferry pinned it against a pier. The tetanus he contracted killed him six weeks later. His son, Washington, took over and spent fourteen years completing what would be seen as both an engineering marvel and an exquisitely beautiful structure.

## A Fatal Allure

The structures that were rising to new heights in the nineteenth century, and that would continue to rise in the twentieth, were to have an effect that could not have been guessed by their engineers. Their height was frightening, but also strangely alluring. In 1937, the Golden Gate Bridge—surpassing the Brooklyn Bridge and others as the longest suspension bridge in the world—was stretched across San Francisco Bay, using the design pioneered by Roebling and made with cables from his sons' company. The Eiffel Tower remained the highest open-air public structure known. Both became immensely popular tourist attractions at the moment of their creation. Within a few decades, they also became the two most popular places in the world to commit suicide.

There is no single profile for the kind of people who are likely to jump to their death. It can be said that wherever there are high structures, there will be people willing to jump from them. Dozens of leaps appear to be spontaneous, like one that took place in Delaware in 1965: two men meet in a bar, and after it closes they decide to drive across the Delaware Memorial Bridge to resume their drinking in New Jersey. When they're stopped by heavy traffic, one dashes from the car, clears the rail, and is gone. On the Golden Gate, one man who went over left this note: "Absolutely no reason except I have a toothache."[4] Studies have shown that for many who leap, the suicidal urge was momentary. One survey of about five hundred would-be jumpers who were talked down from the Golden Gate Bridge revealed that only 10 percent commited suicide later.[5]

For the impulsive, falling is chosen as a means of death for no particular reason other than that it's convenient. Statistics confirm that when people want to kill themselves, they use what's handy. In the United Kingdom in the 1940s and 1950s, when coke gas ovens were common, they were used for nearly half of all suicides. Over the next twenty years, as coke gas was replaced by less toxic natural gas, the suicide rate fell in almost direct proportion to the shrinking number of coke gas ovens. In America, the majority of people who kill themselves—about 57 percent—use a gun; a testimony to how armed we are.[6] In 1999, jumping from a high place accounted for only 2.4 percent (693 people)[7] of all suicides in the United States. Although jumping will probably always account for a small portion of the overall suicide rate, it's clear that where tall structures are common, more people will use them to die. In a study of the five counties of New York, researchers found that fatal jumps occurred most often in the two counties with tall buildings.[8]

While the method may not make much difference to an impulsive victim, it does to the rest of us. Although death by falling represents a small fraction of all suicides, it is an act that grips our emotions with power out of proportion to the numbers, because falling carries a peculiar poignancy. The leapers' time in the air allows reflection for a few seconds, but time enough for—what? What kind of thoughts go through your mind when you see death rushing toward you so fast? We can't help but wonder.

It's a moment that isn't present in other kinds of suicide. Shooting yourself, for example, doesn't have any time between the act and its result. Pull the trigger and you receive the shot. With an overdose there is an unknown gap: time, perhaps, to be discovered; time for a doctor to be called, a stomach pumped; or time for death to come as sleep. But when someone chooses to take a lethal jump, falling's inherent dramatic structure—the leap, the fall, the impact—turns their death into a three-act tragedy. It is the dark version of the story presented by gravity daredevils who conquer nature in the second act to land safely for a triumphant third-act resolution. For suicide victims, the second act functions as it does in any tragedy: it reveals that the victim's moral failure will lead to a horrible end—and that nothing can be done to stop it. What's more, its meaning betrays the promise of progress, by featuring a protagonist who willingly submits to nature's power to kill.

It is also a tragedy that often has an audience, and this is another reason for its empathetic power. The most popular suicide methods are private, the bodies eventually found sprawled in a lonely bedroom somewhere with a stomach full of pills or a shot to the head; we can only guess at the details of the victim's final desperation. In contrast, public jumpers reveal the full drama of their demise.

## The Most Popular Place to Die

That drama can have peculiar effects on whole communities. One example is a case that happened in Pasadena, California, during the height of the Great Depression, just a block from the corner where today's Tournament of Roses Parade is televised.

On a spring morning in 1937, two men were driving under the Colorado Street Bridge when they heard a thud off to the side. Apparently something had fallen from the bridge. They looked up and saw a woman climbing over the balustrade. She jumped from the bridge, 150 feet, and disappeared in the brush. The men scrambled over a fence and pushed through the thicket into a clearing, where they found a weeping child crawling toward the body of a woman not much older than twenty. Other witnesses told the police they had seen the woman lift the little girl above the railing and throw her over. Mother and daughter were taken to a local hospital; the mother lingered two hours before dying, while her daughter, incredibly, was released to her grandmother after being treated for nothing more than bruises and scratches.

Suicides at the bridge were nothing new. The number had increased every year since its opening in 1913, and there had been several inconclusive public debates about how to stop them. But now, the image of a mother throwing her daughter from the railing—and the realization that only luck had saved the girl—incensed the public and brought crowds to the city council's meetings demanding that a suicide barrier be built. Others appeared who were just as strident that a barrier not be built because it would ruin the beauty of the city's architecturally significant bridge. They blamed "out-of-town newspapers" for blowing the problem out of proportion with sensational stories about the tragedy. Their theory was that these newspapers wanted to see Pasadena humbled because it was a winter resort for wealthy easterners. Some suggested curious alternatives: planting more trees in the ravine so that jumpers could see that they would only be maimed; posting a suicide patrol dressed up as ice cream vendors so as not to alarm the tourists. In the end, the drama of the case impelled the city council to order the fence, and it was built in record time.

In communities with high structures more famous than Pasadena's bridge, there's often greater pressure to simply live with the problem of suicidal jumpers, no matter how abhorrent, rather than alter the city's landmark with a barrier. The Golden Gate Bridge is one of these unaltered structures, and as a result, it's the most popular place on earth to commit suicide. Officially, more than a thousand people have jumped. Two or three people a week are talked out of the act by a seventeen-member suicide patrol, but that patrol only works during the day. If a body isn't recovered—and they often aren't—the only way of knowing if someone has gone over is if they were seen or if they left evidence on the bridge—an abandoned car, a pile of clothes, a note. According to the staff at San Francisco Suicide Prevention, the number of jumpers is probably at least two thousand.[9]

With a figure like that, the Golden Gate's popularity for suicide can't be explained as simply a result of it being convenient for impulsive victims, and this fact reveals another side of jumpers. For many, the method and place are extraordinarily important; they are known to select their sites carefully, usually for their beauty. In San Francisco, statistics show that the Oakland Bay Bridge is spurned; virtually everyone jumps from the more appealing Golden Gate, even if they have to drive over the Bay Bridge to get there. Ninety percent jump from the side that faces the spectacular view of the city. In one study, ten people who had survived the leap were interviewed, and all said they had planned to kill themselves on the Golden Gate Bridge and nowhere else—and that if they had been stopped, they wouldn't have tried suicide by another method. Follow-up research showed that they didn't attempt suicide later.[10]

For these people, suicide by jumping is a form of death in which aesthetics matter. In dozens of their stories, their behavior seems a kind of foreplay to their tryst with falling, like the preparations of an apprehensive lover planning to act out a private desire in a public place. They often walk with eyes averted out on the bridge, then back, then out again. One woman caught the attention of a policeman because she had gone through the turnstile three times, but she had a camera; a tourist, he guessed. She finally jumped; the camera was empty.

Sometimes a spouse must be tricked into helping; one woman recuperating from nasal surgery suggested to her husband that they take a relaxing drive on the bridge. When she asked him to stop the car, saying that the motion was causing a nosebleed, she jumped out and went over.

Sometimes there is exhibitionism. "I was walking on the west side with my wife and five children when I first noticed him on the opposite side," one man said of a suicide he witnessed. "He was taking off his coat, then putting it on again, then taking it off—to make sure he would catch my eye. He did that three times. When he was sure I was looking, he folded it neatly and over he went."[11] Sometimes there is modesty; one woman in a tight skirt couldn't figure out how to straddle the rail without exposing herself, which gave a policeman enough time to stop her.

Jumpers often remove garments—their coat, shoes, handbag—sometimes placing them in a neat pile. It seems important to many of them how their bodies meet the air, and they have been seen somersaulting or diving gracefully. "He dove forward with arms outstretched and seemed to glide down," a witness said of a jumper. One man leapt with his hat on. "It was strange," an onlooker reported. "He was holding his hat on his head with both hands all the way down. He was holding onto it like he was afraid it was going to blow off."[12]

There is also a case of a woman who was seduced by falling, or said she was. Her name was Cornelia Van Ireland, and in 1941 she was twenty-two years old and engaged to be married. One afternoon she skipped work and took a walk on the Golden Gate Bridge. She spent some time staring at the water, and although there is no evidence that she was depressed or wanted to die, she vaulted the short fence and sailed out into the air. Here is what she said from her hospital bed: "I don't know what happened. I had an irresistible impulse to jump, and suddenly I clambered over the railing and fell into space. I had no particular sensation going down. I know I prayed, but I had no feeling of pressure against me, no sensation of falling. I don't remember when I hit the water, but I know I was conscious. I was conscious every moment."[13] She had fallen two hundred and fifty feet and hit the water at eighty miles an hour, breaking both arms, fracturing several vertebrae, and tearing the flesh on her legs and thighs. She survived only because two men in a boat happened to be nearby and pulled her out within minutes.

## The Aesthetics of Death

Unlike any other form of suicide, falling to one's death is tangled up with aesthetics, both private and public. Those who jump are attracted to beautiful structures; those in charge don't want to mar the structure's beauty with a suicide barrier. There are other considerations, of course: the expense of a barrier and the question of whether people should be protected from themselves. But the controversy boils down to what's more valuable: lives or beauty. And surprisingly, it's not an easy choice.

Suicide barriers do prevent suicides, but they also interrupt graceful design and more significantly, destroy the pleasure of an unobstructed view from a high place. The very sensation of feeling free at heights—the same sensation that attracts suicides—is also a precious experience for the rest of us. If we decide that preventing a relatively small number of people from killing themselves is more important, how long will it be before Glacier Point in Yosemite or the rim of the Grand Canyon have barbwire-topped fences strung along their edges? If we decide instead that these places should be kept clear, then why not leave the Golden Gate Bridge alone as well?

To date, we have taken a middle ground, sometimes willing to sacrifice the freedom of height for public safety, sometimes willing to pay the price. In Pasadena, the case of the mother who threw her child over the edge motivated the community to build a barrier. In San Francisco, the city has been presented with the same horrifying incentive more than once.

On a summer day in 1945, August De Mont and his daughter walked to the railing of the Golden Gate Bridge. De Mont lifted five-year-old Marilyn over the rail and set the girl on the girder just below. He told her to jump. She did. Her father followed her down. Unlike the case in Pasadena, both died. In 1993, it happened again: A distraught man appeared on the Golden Gate with a bundle in his arms. When bridge officials approached, he ran to the railing and tossed the bundle over—

which the officials suddenly realized was a child. He then jumped himself.

Before both cases, San Franciscans had been arguing about suicide prevention on the bridge for years, but unlike Pasadena, the wrangling went on in spite of these unusually horrific events. Topping the guardrails with an electric fence or barbed wire was first suggested in 1951 but rejected because it would be dangerous to maintenance workers who had to go over the rails every day. In 1953 it was proposed that the rail simply be raised from four feet to seven feet; this time the idea was defeated on the grounds that a tall fence wouldn't stop anyone bent on suicide—an indefensible conclusion. When the seven-foot fence went up on the Colorado Street Bridge in Pasadena, the number of deaths, which had climbed to nearly eighty by the time of the mother and daughter's fall, dropped to zero the next year, and only twelve people have managed to climb over it since. At bridges in Massachusetts, the Eiffel Tower, and the Empire State Building, barriers produced similar results.

As of this writing, San Francisco is again in the throes of controversy about a proposed barrier on the Golden Gate. In 1998, bridge officials set up a prototype in a parking lot and invited comments and testing. The barrier consisted of cables running horizontally about four inches apart, a design that was meant to obstruct the view from the bridge as little as possible. The testing didn't go so well: five people were able to climb over the barrier as well as slip between the wires. An architectural advisory panel didn't like its looks, and neither did Supervisor Barbara Kaufman, who said it looked like "barbed wire at concentration camps."[14] The *San Francisco Chronicle* also hated the fence, opining that "to add such a jerry-rigged monstrosity to the graceful span would be a grave insult to San Francisco's landmark masterpiece,"[15] although the editorial did agree that a barrier of some kind was needed.

To date, the rail on the Golden Gate bridge remains only four feet high.

Karl Wallenda, the last of the traditional gravity heroes, walks the wire above Busch Memorial Stadium in St. Louis in 1971, at the age of sixty-seven. In his later years Wallenda typically performed alone; he fell from the wire to his death in 1978. *(Corbis)*

CHAPTER 5

# Fall of the Noble Daredevil

I am finished. —THE LAST WORDS OF TONI KURZ, CHALLENGER OF THE NORTH FACE OF THE EIGER

AS THE NINETEENTH century closed and the twentieth opened, the old image of the gravity hero began to fray at the edges. At its peak, in midcentury, this picture of a noble individual triumphing over nature had propelled Whymper, Léotard, and Blondin to the heights of acclaim. But shortly after, perceptions changed. Whymper had been party to a tragic accident, the trapeze act Léotard invented had lost its novelty, and the Niagara wirewalkers who followed Blondin seemed to simply confirm the stunt's ease. And on an autumn afternoon in 1901, the image was degraded even further, when the first person to attempt Niagara's most dangerous and anticipated stunt—going over the Falls in a barrel—turned out not to be a brawny adventurer but a plump, sixty-three-year-old widow named Annie Taylor.

She wasn't a man and didn't subscribe to manly motivations. She didn't compare herself to great generals, as Sam Patch had, and didn't talk about conquest, as mountaineers still did. Nor did she see herself as an iconoclast—as a champion of women, say. She just wanted money.

Annie, a farmer's daughter who had grown up in Auburn, New York, was a fiercely independent woman who had been on her own since her husband died in the Civil War. She had gone on to get a degree in education and in 1870 traveled to the frontier town of San Antonio, Texas, where she taught for a year before being promoted to vice principal. Evidently restless, she left three years later, trained as a dance instructor in New York, and then taught in cities around the country, in Chattanooga, Birmingham, and San Francisco. She toured Europe, compliments of a wealthy friend. She opened a dancing school in Bay City, Michigan, but it failed, so she took to the road again, to Mexico City, El Paso, St. Louis, Chicago, and finally, back to Bay City. As she aged, it became harder to get students.

By 1901, she was living off the charity of relatives, a humiliating downturn for such a proud woman. But there was no way for her to earn money for retirement in Victorian America, except as a domestic. Visions of spending her declining years in a poorhouse loomed. "I didn't want to lower my social standard," she wrote later, "for I have always associated with the best class of people, the cultivated and refined. To hold my place in that world I needed money, but how to get it?"[1] She was reading a newspaper story about the crowds at Niagara Falls and the trips others had made in barrels through the whirlpool, when, incredibly, she decided that going over the Falls in a barrel would be her best option for financial independence.

Alone in her rented room, she cut out pieces of cardboard and bound them together into a full-size model of the barrel she wanted. Then she called a cooper to make it, telling him to come in through the back door so no one would think she was entertaining a man in her room. He thought the old woman's plan was outrageous, and refused to build the barrel. She talked him into it and picked out all the lumber herself. She hired a manager, Frank Russell, who had promoted high-diving acts, and told him the fiction that she needed money to pay off the mortgage on her ranch in Texas. She also lied about her age, saying she was forty-two.

In October 1901, only three months after she had hatched her plan, she was ready. She sent Russell to Niagara first, to announce to the press that his client, whose name he wouldn't yet reveal, would be the first person in history to go over the Falls and live. He said she had "scaled the Alps, made dangerous swimming trips, and explored wild, unknown countries," among other fabrications.[2]

Annie arrived at Niagara Falls two days after Russell, on October 13, surprising reporters: she hardly looked like the forty-two-year-old adventuress Russell had described. She was short, ample, and looked all of her sixty-three years.

Although Russell was supposed to figure out a way to make money from Annie's stunt, he had no plan beyond booking Annie into lecture halls after she made her headlines. Unlike Farini, he failed to make arrangements with the railroads to sell seats on the train to view the performance. Nor did he have bleachers set up for those willing to pay. Annie's only profit the day of the event would come from selling photographs of herself. Anyone could watch Annie risk her life for free.

She first tested her barrel by sending it over with a cat inside, although reports are mixed as to whether the animal survived. Then Russell announced that Annie would perform her historic stunt on Sunday, October 20. Hundreds of people came to watch, but Annie never showed, and Russell had trouble explaining why (most likely, the photographs she had planned to sell weren't ready yet). Later, Annie announced that she would go on Wednesday, the twenty-third. This time blustery winds stopped her; she swore she would come back the next day.

And she did. From the American shore she stepped into a skiff with her barrel and her two assistants, rivermen John Truesdale and Billy Holleran. The three rowed a short way to Grass Island, where they beached and prepared for her launch toward Horseshoe Falls.

After removing her broad black hat and black jacket, Annie, in a modest skirt, stockings, and blouse, climbed into the four-foot barrel, strapped herself into a harness, and cushioned herself with a small mattress. She gripped the two handles inside. Her assistants sealed her in with a two-inch-thick cover. Annie was worried about suffocating (this had happened to a woman who had taken a sealed barrel through the whirlpool below the Falls) so she had Holleran spend twenty minutes pumping air into her barrel with a bicycle pump—a futile effort, as it turned out, because the barrel was so permeable that it started leaking the moment it was put in the water. "How much water is in it?" Truesdale asked, as she bobbed in the shallows at the edge of the river. "About a pailful,"[3] she answered. Truesdale told her it wouldn't hurt her, and let her go.

Inside the dark barrel she heard the river gather speed as she was drawn toward the center of the channel. She heard rapids, then was in them, the barrel flipping and turning, knocking her against the wood. Then she felt herself fly out into the air—but this, she knew, was just the forty-foot drop a half mile above the Falls. When she hit, the barrel was shoved underwater with such force it scraped along the riverbed. Then it rose, back into the cauldron.

In another moment, everything calmed down. The barrel stopped spinning, and the hundred-pound anvil in its base pulled the barrel upright. She sat still, gripped the straps, and tried to relax as she picked up speed and heard the roar coming closer. Now it was deafening. The barrel began to tip, then slide; suddenly she was shot out over the 165-foot drop, flipping wildly, end over end through the air.

Inside she was beaten against the staves. She braced against the impact of landing, but it never came. The roar filling her ears suddenly hushed as she felt the barrel sucked underwater, where all sound ceased. Then the barrel began to rise. She shot up out of the water (fifteen feet, witnesses said), heard the thunderous cataract, slammed back into the river, bobbed up to be launched into the Falls themselves, and then was thrown into the broken rocks near the edge of the Falls.

She became dimly aware that someone was pulling at the barrel, and heard men shouting orders at each other. There was pounding and prying, and then daylight—but she was too exhausted to get out. Truesdale took a saw to the barrel, cutting enough of it away to drag Annie from it.

Her head was cut open, she was scraped and bruised, but otherwise uninjured.



IT HAD BEEN just over a decade since Steve Brodie had made a fortune from his claim that he had jumped from the Brooklyn Bridge. How much more famous would Annie be? She had pulled off the one gravity trick that had daunted every male daredevil up to that point. If history was any guide, crowds should have hoisted Annie on their shoulders, clamored for her autographed pictures, and begged for her to speak at venues from coast to coast.

The crowds did come—at first. Annie traveled to appear at the Pan American Exposition in Buffalo, a wonder of the age, with its nearly four-hundred-foot-tall Electric Tower (lit by power drawn from Niagara Falls) and sprawling grounds filled with domes, minarets, statues, and facsimiles of exotic lands: there was a Japanese village, an African village, Cleopatra's temple, and Venice in America, "a perfect replica of the living city, with palaces, shops, bridges and canals, gondolas and gondoliers,"[4] as the exposition's program put it. As she stood for six hours in a receiving line, Annie must have thought her dream had come true. "There was a continuous stream of humanity," noted the *Buffalo Evening News*, "and Sergeant Smith Jackson and his squad of Exposition guards had their hands full keeping the lines straightened out and preventing crushes."[5]

But it was the last day of the exposition, and there was nowhere to go from there. "If she had been a beautiful girl," her manager grumbled, "why we would have made thousands."[6] Only newspapers around Niagara lauded Annie, using the now tiresome hyperbole that had been customary for gravity daredevils of the preceding century. "The climax of Niagara wonders,"[7] opined the Buffalo *Courier*. The Niagara *Gazette* said Annie had achieved the "foremost rank in the list of those who have dared to toy with Nature."[8] But this was just local boosterism. Elsewhere in the country, Annie's stunt was minor news, if it was noted at all.

Annie had imagined herself on a lecture circuit after her accomplishment, perhaps speaking before men in tuxedoes and ladies in evening wear. Instead, the only booking after the exposition that Russell could muster was at a dime museum in New York. Horrified, Annie turned it down. When it became clear that people weren't clamoring to see her, she finally agreed to pose with her barrel in store windows in cities across the Midwest—but this brought in little money. Russell was frustrated, and his solution was to leave Annie and take her barrel with him.

Russell sold the barrel to a theater company that was mounting a play about Annie, for which she would get nothing. Annie sued the company to get her barrel back, then retained a new manager, William A. Banks. After a few appearances at state fairs, he stole the barrel too, and all her pamphlets, and hired a young beauty to pass herself off as Annie. Meanwhile, Truesdale and Holleran, Annie's former assistants, made a movie of an empty barrel going over the Falls, starring themselves, and sold it to the nickelodeon houses as a record of Annie's plunge.

For the next twenty years, Annie would haunt the sidewalks of Niagara Falls City, an old, wrinkled woman sitting next to a replica of her barrel, selling postcards and her autobiography. In 1921, at the age of eighty-three—but now claiming she was fifty-seven—she entered the county poorhouse. "It is quite a change for me to come here when I have been used to being entertained in senators' homes in Washington and traveling extensively," she said, "but I feel that it is no disgrace and if all my plans materialize, I shall not remain here long."[9] She died there two months later.

IN SPITE OF being the first to perform the ultimate Niagara stunt, Annie never got the respect that had been lavished on other gravity heroes. If she had been a man, it might have been different. But there is more to the story than sexism. In visual terms, the wirewalkers and high divers of earlier times presented a more accessible image of heroism: lone figures standing in the free air high above the river, they invited the spectators' empathy. Their motivation was something men could understand

and women had to admire, at least to some degree: to test one's ability in a high-stakes crucible that could lead to personal honor as well as wealth and fame.

As a barrel rider, on the other hand, Annie was the picture of isolation and masochism—not so much a challenger to gravity as a punching bag willing to be beaten senseless by it. Instead of the commanding image of a man standing on the lip of a great precipice, in control of his fate, Annie, once set adrift in her capsule, was obviously helpless, just another piece of flotsam at the mercy of the river. And what normal woman—or man, for that matter—could find in their own psyche anything similar to the desire that drove Annie?

After Annie, virtually all daredevils who came to Niagara aimed at going over the Falls in some contraption or another, and virtually all of them were considered eccentrics, not heroes, and reasonably so. The first man to go over, ten years after Annie, was saloon-owner Bobby Leach, who in 1911 crawled from his steel keg dead drunk, shouting, "Ain't nobody got anything on me now!"[10] Several more went over after Leach; most lived, a few didn't. One who didn't was Charles Stephens, a fifty-eight-year-old barber, parachutist, and high diver who tried in 1920. In spite of being warned several times that his barrel was too badly built to make the trip, he went anyway. His barrel was taken apart by the Falls. Later, searchers were able to piece together what happened. Stephens had tied his feet to the anvil that weighted his barrel. The cataract had apparently ripped off the barrel's bottom and taken him with it—at least most of him. All that was found was his right arm, still strapped to his harness. The tattoo on it, bearing his wife's name, read, "Forget me not, Annie."

## Death Wish

After decades of gravity daredevils who had been taken seriously as public heroes, events at Niagara showed that many who were willing to fling themselves into the void weren't necessarily brave but were more likely mentally unbalanced. And if this were true of Niagara stunters, could it also be true of other gravity challengers?

Now, at the beginning of the twentieth century, there appeared a new scientific theory that seemed to answer yes. According to psychoanalysis, in a mysterious part of the brain called the subconscious, Niagara barrel riders—and perhaps high divers, wirewalkers, mountaineers, and anyone else who risked a fatal fall—actually wanted to kill themselves.

The idea of a death wish comes from Freud, who in 1923 postulated that there were only two basic human instincts: one pushing us toward procreation, the other urging us to die. Freud saw the battle between these two instincts as going on inside the most primitive (and unconscious) part of the mind, the id. Another part, the conscious ego, tries to control the id's demands so that sexual depravity or suicide don't result. The third part of the mind, the super-ego, makes judgements about how well the ego is doing its job.

According to Freud, the healthy ego diffuses the id's demands by channeling them into what seems like unrelated behavior; the familiar example is the hard-charging executive who expends his energy in the boardroom instead of the bedroom. The neurotic ego, on the other hand, tries to prove it has mastery over the death instinct by actively courting death and then defying it.

Freud was sixty-seven when he formulated this theory, and some have speculated that it was his own obsession with death (he had thought of himself as decrepit and aging for two decades) that accounts for it. But whatever its origins, the idea of a death wish confirmed what some people had long suspected: that mountain climbers, high divers, and all the rest were crazy.

In the literal act of falling, Freud saw an unconscious desire of "the giving up of bodily equilibrium," which was an expression of sexual fantasies. So, too, were dreams of falling. "Their interpretation, when they occur in women, offers no difficulty, because they nearly always accept the symbolic meaning of falling, which is a circumlocution for giving way to an erotic temptation."[11] From a Freudian point of view, then, a gravity daredevil might be described as a neurotic driven by both sexual and death instincts at the same time. Whatever was left of the idea that defying gravity was heroic was beginning to come apart at the seams.



## The Extremists

Thirteen years after Freud hatched his death-wish theory, a mountaineering accident on a particularly vicious Swiss peak would raise questions about the sanity of some climbers, even among the sport's defenders. Years before, in 1865, the tragedy on the Matterhorn had taken four lives and ended mountaineering's Golden Age, yet the sport had gone on to become more popular than ever. By the early twentieth century, thousands of ordinary Europeans were now climbing. And following the early example set by Hudson and Kennedy's guideless climb of Mont Blanc, they were now climbing by themselves. Uninhibited by the conventional methods a guide would impose, some of these amateurs were pushing far beyond what any guide would dare, earning them the moniker *extremists* among the mountaineering establishment.

One of this new breed was a young German soldier named Toni Kurz. In 1936, at the age of twenty-three, Kurz, along with his partner Anderl Hinterstoisser, went to Switzerland to climb the north face of a mountain whose name is German for "ogre"—the Eiger. The sides of the Eiger had been climbed years before; in 1867 John Tyndall climbed the west flank, repeating the route of the first ascent made by Charles Barrington in 1858. Looking down the nine-thousand-foot drop, he said, "A wilder precipice is hardly to be seen than this wall"[12]—and considering his mountaineering record, he was someone who would know. Other routes along the mountain's flanks and ridges were found and also scaled; it was the tremendous wall that remained unclimbed and terrified traditional mountaineers.

The mountain is shaped like a tall pyramid, with one side scooped away, a crumbling, concave rock face interrupted by stretches of sheer cliffs and steep ice fields that jut out like enormous tilted shelves. The weather is horrible. Storms come onto the Eiger in minutes, blowing at a hundred miles an hour, sending rain or snow or hailstones down the face, even when it is sunny and warm in the village below. The rain is funneled into torrents that flood through rock chimneys, huge waterfalls that surge over climbers. The rain freezes to coat the rock with a slick sheet of ice, thick enough to hide every handhold but too thin to allow a boot's crampons to grip. The rain seeps into the Eiger's cracks, freezes, expands, and breaks the rotten limestone into chunks that sail out from the face like cannon shots. Sometimes they hit a shelf and bring down an avalanche or shards of ice the size of windowpanes.

It was the kind of climb that the Alpine establishment believed shouldn't even be considered. "One cannot help regarding the contemplated climbing attempts on the North Face of the Eiger with serious misgivings," the chief guide of Grindelwald had written. "They are a plain indication of the great change which has taken place in the conception of the sport of mountaineering."[13] By "great change" he meant the rise of extremists like Kurz and Hinterstoisser. This new kind of mountaineer shocked traditionalists with their technique of pounding metal spikes into the rock, threading a rope through, and pulling themselves up. To the mountaineering establishment this was heresy: they insisted that climbers should use only ice axes to climb a sheer cliff, and that rope should only be used for safety, not for climbing itself. With their radical approach, the extremists aimed precisely at climbs that horrified the guides, climbs like the Eiger's north face. The extremists were resented by the Alpine guides in the valley beneath the Eiger, and in turn, the extremists didn't think much of the guides. The English, too, disliked the extremists, because of their methods, because they were not of the upper class, and because they were almost exclusively from Axis countries. They were seen as subverters not only of the mountaineer's imagined code but also of civilization itself. Many believed that

Hitler had ordered Kurz and Hinterstoisser to conquer the north wall to prove the superiority of the Aryan race. In fact, Hitler had not, and the superior officer of Kurtz and Hinterstoisser's regiment, after discovering their plan, had sent orders to Grindelwald for them to stop, although these arrived too late.

A year earlier, two extremists had been the first to try the Eiger's north face. Like Kurz and Hinterstoisser, they were also German. When they died, one falling off the face, the other frozen in his stance at Death Bivouac, the mountaineering establishment had little sympathy. The Swiss journal *Sport* called the attempted climb "mountaineering perverted into monkey tricks" and in a nod to Freud, declared that "it is a German's affair if he is compelled to exercise his modern psychology in 'direct' ascents." In other words, trying to climb straight up a cliff is madness. "Even had the two Munich scramblers succeeded in their attempt, it would still have been a mockery of classic Alpine climbing, an evil demonstration on Swiss peaks."[14] The *Alpine Journal* went on to call the climb, "a flagrant example of the neglect of every sane principle of mountaineering in an attempt to gain cheap notoriety by accomplishing mechanized variants to former routes."[15]

It was old criticism in a new bottle. Like the invective hurled over the decades at Garnerin, Patch, and Blondin, the charge was immorality—the climb would be an "evil demonstration." But picking up on Freud's new idea, the climbers were also accused of literally being psychotic: guilty of neglecting "every sane principle."

The *Alpine Journal*'s most revealing assertion, though, is obscured with jargon and lies within the phrase "mechanized variants to former routes." The Eiger's summit had been reached before, by climbing its flanks, and this was well and good. But pursuing a "mechanized variant," a route that required the machinery of pitons and carabiners, was "evil" and insane. The primary difference between the two routes was that the extremists' route increased the chance of falling.

Apart from unkind words, this schism in mountaineering created a much more serious problem for the extremists: if they got into trouble, the local guides couldn't help. "We should find it impossible to force our guides to take a compulsory part in the kind of acrobatics which others are undertaking voluntarily," warned the chief guide of Grindelwald. "No one can expect the dispatch of guides, in unfavorable conditions, on a rescue operation"[16]

When Kurz and Hinterstoisser arrived in the valley, it was already filled with newspaper reporters who smelled blood, because they weren't the only extremists that season who wanted to try the north face after the deaths the year before. There was even a woman, Loulou Boulaz, who climbed up two thousand feet before she turned back.

The German pair were only in the valley a short time before meeting Edi Rainer and Willy Angerer, two Austrians who also wanted the north wall and who had already climbed it up to the Rote Fluh, a thousand-foot vertical stretch that seemed unclimbable. After talking, the four decided their chances would be better if they all roped up together.

They set out on a summer's night at two in the morning. The weather was good, and they reached the Rote Fluh by nine. Here, Hinterstoisser suggested that instead of going up the sheer cliff, they go around it. He worked his way sideways for 130 feet across the cliff—a technique known as the traverse, invented by the new climbers—hammering in pitons on which he hung a rope to be used as line for the others. In the years to come, the route, known as the Hinterstoisser traverse, would be called brilliant and would be used by virtually all the north-face climbers who followed.

And here, Hinterstoisser made a crucial mistake: he pulled the rope from the traverse once they had finished with it. They didn't expect to need it; their plan was to reach the summit and walk down the mountain's easy flank.

They were on the First Icefield when trouble came. Unprotected from above, the climbers cringed with each step as rocks and ice chunks rained down. They climbed fast but couldn't escape before a stone shot off from the huge rotten wall above and struck Angerer in the head. By the time they reached the top of the ice field, Angerer was staggering along only with Rainer's help. Still, they continued to climb. At the bottom of the Second Icefield they dragged themselves into a cave for the night.

The next day they moved out onto the vast white ice field, chopping out each step across its thousand-foot length to the Flatiron, a terrible, overhanging rock. Here, they camped again, as more bad luck appeared: clouds were rolling in.

They had barely started the next morning, with Hinterstoisser and Kurz on a separate rope in the lead and far ahead of Rainer and Angerer, when all four stopped for a half hour. To observers in the valley, it looked as if Angerer was too injured to go on and that they were trying to decide what to do. Finally, Hinterstoisser and Kurz climbed back down to the other two, and they all began their retreat.

As they did, the wind began to rise and the clouds grew dark. They worked their way back across the Second Icefield and were descending to the First Icefield when it began to mist, then to shower. Wet and utterly spent, they camped.

In the morning, Angerer was worse, judging from their slow descent. They crawled down the First Icefield, the rain now freezing on their soaked jackets, down to the traverse, where they had not left a rope.

Hinterstoisser was determined to cross again and tried, but the rock was now a sheet of ice. He tried for hours. When he was exhausted, Kurz tried. The rain turned to snow. Angerer sat, not moving. They were left with no choice but to try to go straight down—into what, they couldn't have known, although they may have been trying to reach the window of the train tunnel that ran through the mountain's heart. This was a few hundred feet below and to the west.

Rainer hammered in a spike and clipped his harness to the cliff; he then wrapped the rope around his body, ready to catch the others if they fell as they descended. Roped together, Hinterstoisser led, followed by the ailing Angerer and then Kurz.

The retreat was going as planned, and in spite of the rain, it looked as if they might succeed. Below, a railway man in the train tunnel threw open a window and called to them. "We're on the way down," he heard Hinterstoisser, the lowest man on the rope, reply. "All's well!"[17]

A short while later they all stopped climbing momentarily, to rest on their holds. Rainer, the top man, drove a piton into the face, clipped the rope from below through the attached carabiner, and then fixed it to his harness to secure the three climbers strung out below him. Hinterstoisser, at the bottom, decided to unclip from his rope to give the others a little slack to work with—and in that moment, the tension suddenly gone, his muscles drained, his sense of balance dull from fatigue, Hinterstoisser fell, tumbling away, down the huge wall.

Angerer, just above, caught in the middle of trying to adjust, lurched back and slipped off his stance, with the rope wrapped around his body and neck in a stranglehold. His weight on the rope yanked Kurz from the cliff, sending him flying into the air in his sling, and above, it jerked Rainer hard against the anchoring piton and pinned him there. The rain turned to sleet.

It was only an hour or so later when the railway man called again from the window. By this time Angerer had apparently choked to death and Rainer had died of exposure; Kurz now desperately cried out, "Help! The others are all dead. I am the only one alive!"

It was nearly nightfall before rescuers could be brought in. They climbed out through the window and got close enough to shout a conversation with Kurz, a hundred feet or so above them. He said he had no rope or pitons, only his ice ax; Hinterstoisser had been carrying the hardware. The sun was setting. He begged them to climb up, but these were Alpine guides, and they had no idea how to climb the overhang that cut them off from Kurz. They said they would return at dawn's light. "No! No!" Kurz screamed, "I'll freeze to death here! Don't leave me!"

Kurz spent the night hanging in a storm between the corpses of his friends, twisting in the black blizzard, hearing rock and ice rumbling down around him. He lost the glove on his left hand, and during the night his arm froze solid.

At daybreak the rescuers returned. They climbed up below the overhang and then told Kurz they could climb no further. Kurz urged them to take the route that the team had, up and across the traverse; from there they could drop a rope to him. It was impossible for them. They had another plan, one that would require a superhuman effort from Kurz. He had little choice but to agree, and followed their instructions.

Kurz worked his way down the rope, gripping it only with his right hand and legs. When he reached Angerer and cut the rope, the body didn't fall: it had frozen to the rock. He then began the climb up, wrapping the rope around his body, swinging his ice ax, dragging up a bit, and rewrapping the rope. It took him several hours to climb forty feet.

When he got to Rainer, he cut the rope from him. Then, following the plan, he took the frozen rope in his teeth and began to work apart its three strands. It took him five hours. Next he tied the end of each strand to the next, using just his teeth and his one good hand, to make a 120-foot line. He tied a stone to the end and lowered it.

The men below fixed a fresh hundred-foot rope to the line, along with a hammer and carabiners, and Kurz pulled it up. But the men started shouting. The rope they had attached was too short; they saw the end disappear beyond the overhang above them as Kurz pulled it up. He lowered the rope again, and to its end they tied another hundred-foot rope.

Kurz pulled it up. Next he wedged a piton into the rock and hammered it in, his weak blows coming sporadically, only after he could gather the strength to lift his right arm. Suddenly, they screamed below him. Angerer's frozen body had fallen away, released from the cliff, and at first they had thought it was Kurz. "I am still here," Kurz cried out.

He tied the rope's end to the piton, snapped a carabiner to his sling and passed the rope through it. He wrapped the rope around his body and clenched it under his left arm, then slowly began backing down the cliff.

Finally, he reached the overhang that separated him from his rescuers, and carefully stepped backward, down its slope. He was very close now and could hear them shouting encouragement. To go over the bulge, he laid down on the rock. The men below saw his boots appear; keep coming, they shouted, keep coming. He was only fifteen feet above them.

Then he stopped. The rope above was taut. Kurz looked at where it passed through the carabiner at his waist. The knot that tied the two ropes together was too big to slip through.

"Force it through," they shouted. Kurz said it wouldn't go. He bent over and chewed on the knot. The exhausted, desperate man chewed on it and it would not go through. The rescuers tried again to climb to him, and got close enough to tap the spikes of his boots, but no farther.

Kurz then said something. "What?" the men asked. "*Ich kahn nicht mehr*," he said, clearly this time. *I am finished.* And that is all they heard.

## An Obsession for the Mentally Deranged

If Toni Kurz did not die at that moment, he died only moments later. The men trying to rescue him had watched as he went limp. Later they cut his body down with a knife lashed to a pole.

When the mass of reporters began to file their stories about the latest deaths on the Eiger, respectable mountaineers redoubled their attack on the extremists, once again using the language of psychology to condemn. "The Eigerwand continues to be an obsession for the mentally deranged of almost every nation," the president of the Alpine Club wrote. "He who first succeeds may rest assured that he has accomplished the most imbecile variant since mountaineering began."[18] The Norwand (north wall) of the Eiger began to be called the Morwand (murder wall).

But the extremists had already changed mountaineering in a crucial way. No longer was getting to the top of a peak the primary goal; getting there the impossible way was. Challenging gravity, more than before, was the new game, and there was nothing the Alpine Club or any other institution could do to stop it: by the late 1930s, nearly a hundred mountaineers a year were dying during climbs.



Among them were those still aiming for the Eiger's north face. A year after the Kurz fiasco, two Austrian climbers, Franz Primas and Bertl Gollackner, went up the west flank to get a look at the wall they planned to climb later. A blizzard pinned them down, Gollackner died of exposure, and Primas was rescued, barely alive. The following year, 1938, Bartolo Sandri and Mario Menti made a good start up the face, but weren't anywhere to be seen the next day by those peering through their telescopes in Grindelwald. A rescue party found their mangled bodies at the cliff's base.

Later that year, Anderl Heckmair, Ludwig Vörg, Heinrich Harrer, and Fritz Kasparek made their assault on the north face. By the third day, they were three-quarters of the way up the wall, to the steep ice field known as the Spider, when the weather turned. They kept on, in spite of the lightning cracking around them, when they heard a whistling sound: an avalanche was coming. It swept down the ice field; Harrer clung desperately to the face as snow poured into the space between him and the mountain, nearly shoving him off the cliff. On the fourth day, Heckmair was leading up a steep gully when a piton pulled out and he fell. Rocketing down the channel, he hit Vörg, knocking him from his stand—but his anchor held. Had it not, the chain reaction—like that on the Matterhorn—would have sent all four men tumbling down. Later on in the day, Heckmair was hacking his way up a wall of ice through a snowstorm when Vörg, just below him, thought he saw rocks beneath the frozen sheet. When the snow cleared a bit, he realized with a shock that he was looking through a *hole* in the ice; the "rocks" were outcroppings thousands of feet below them on the other side of the mountain. Above, Heckmair was anchored to nothing more than a sheet of wind-blown ice. Had it broken, they would have fallen straight down the southern cliff. They climbed down, reconnoitered, and a few hours later reached the summit of the Eiger, the first to climb the dreaded north wall.

In succeeding years, the north face of the Eiger remained a deadly challenge for mountaineers; more than forty people died attempting it. It also became accepted as a legitimate climb, signaling that what was once extreme had become mainstream. By that time, mountaineers would hardly consider a climb worthy unless it involved hanging on a sheer cliff and facing the possibility of a brutally fatal fall.

## A Wirewalker Makes Good

After Robert Cocking's death in 1837, parachuting exhibitions were roundly condemned, and there were often attempts to stop them. In Cheltenham, England, in 1839, public outcry was such that John Hampton had to swear to a gas-works owner that he wouldn't make a descent but would only display himself inside his parachute's basket, dangling from a balloon tethered only thirty feet above the ground—otherwise no gas would be provided to fill his craft. Hampton agreed, but as soon as he was aloft he cut the cord, sailed up to six thousand feet, and floated safely down under canopy to land in Badgerworth. In 1852, Louise Poitevin made a successful drop in Cremorne Pleasure Gardens in London; the next day the police arrived. Citing public outrage, they closed the show.

But for every incensed citizen who called the cops, there were hundreds who willingly paid to see the spectacle—especially on the Continent. Poitevin went on to tour Europe, making a total of thirty-five descents. During the same years in Paris, where parachuting had begun, Eugène Godard made more than fifty descents and became the star of the Hippodrome.

The safety potential for the parachute remained undiscovered: no one could see any real use for it. An 1851 article on ballooning made the point clear:

> Two words against parachutes. In the first place there is no use to which, at present, they can be applied; and, in the second, they are so unsafe as to be likely, in all cases, to cost a life for each descent. In the

concise words of Mr. Green, we should say—"the best parachute is a balloon; the others are bad things to have to deal with."[19]

When Mr. Green (who had been the pilot of Cocking's balloon) said that the best parachute was a balloon, he meant that literally. Balloonists, through accident at first, discovered that a deflated balloon would usually gather in the netting to form a canopy; acting as a parachute, it would provide a safe landing. Some balloonists "ripped" their balloon intentionally to have it form a parachute; in the 1830s, American balloonist John Wise traveled the country making a show of the technique.

Occasionally, ripping didn't work. In the 1880s, German balloonist Hermann Lattemann died trying the stunt. But parachutes seemed no more reliable. In 1854, the spectators at Cremorne Pleasure Gardens saw François Latur fail to get his parachute detached from his balloon; the descending balloon then slammed him into treetops, and he was killed. In 1874, the Cremorne audience saw another fatality when Vincent De Groof's ill-designed parachute shriveled and collapsed.

By the 1880s, it looked like the parachute was at a dead end. It was a daredevil's gimmick and nothing more. But then, in 1885, a new era would arrive, ushered in by a wirewalker and yet another showman, neither of whom knew anything about parachutes.

Tom Baldwin had been on his own since he was twelve, when Rebel soldiers had ridden across the Missouri River to raid his hometown—Quincy, Illinois, a Union stronghold—and shot his parents dead. He had worked at a number of odd jobs. At one, at a sawmill, he had taught himself to walk a tightrope, using piles of sawdust to break his fall. At fifteen, he had landed a job as a professional wirewalker in a traveling circus, and the job had stuck. By 1885, he had been earning his living at it for sixteen years.

Around this time Baldwin was in San Francisco, walking a wire that stretched from the Cliff House restaurant to Seal Rocks, a few boulders seven hundred feet out in the sea. Nearby was Park Van Tassell, who had his own attraction: a tethered balloon that, for a fee, people could ride aloft for a view of the bay.

Van Tassell had seen a picture of a parachute descent in an encyclopedia and thought he would capitalize on the kind of show that had already proved its popularity in Europe. He didn't have a parachute, though, and didn't know where to get one, so he decided to make one himself. Being completely ignorant about parachute design, he didn't bother with the rigid pole-and-strut arrangement that had been the standard since Garnerin's time: he just cut out a round piece of canvas, sewed in a wheel from a horse carriage at the top to provide a vent, attached the chute's lines to a trapeze at the bottom, and was done. Now he suggested to the thirty-one-year-old wirewalker that he could earn some extra cash if he would be the jumper. Baldwin agreed, providing he could become a partner in the enterprise. They tested the chute, then convinced a street-car company to sponsor a drop to take place in Golden Gate Park.

On January 30, 1887, thirty thousand people gathered to watch. The system that Van Tassell had stumbled on was the picture of simplicity; there wasn't even a basket attached to the bottom of the balloon to get in the way. Instead, Baldwin sat on a bar, holding in his lap the trapeze attached to the limp parachute, which was draped from the side of the balloon. A thousand feet above the park, Baldwin slipped off his perch clutching the trapeze. His weight snapped the ties holding the chute; after a short drop, it popped open, and Baldwin floated to safety. It was the first time a soft parachute had ever been used, and it immediately proved that parachutists' concern for keeping a canopy open with a pole and struts was unnecessary. A slack canopy, when it hit the air, would open itself.

While the event was a success, the partnership wasn't. Van Tassell and Baldwin fought over their pay and broke up, but Baldwin kept the parachute. That summer he took his show on the road, jumping at Rockaway Beach, Syracuse, and Kansas City. Not realizing they were seeing the first truly reliable parachute, critics challenged the shows. Respected balloonist Henry Coxwell said the stunt "shows a want of aeronautic common sense."[20] The city fathers of Chicago refused to let Baldwin perform. In Minneapolis, though, he got a break. None other than Buffalo Bill Cody, after seeing Baldwin's act, took him aside and suggested he go to England, where he could hook up with a friend of Cody's who really knew how to put on a gravity show. The friend was Signor Guillermo Antonio Farini.

After his Niagara success, Farini, as mentioned earlier, had produced two phenomenally profitable trapeze acts in England: his son, El Niño, and Lulu, also his son, disguised as a woman. In 1888, he had been hired to bolster attendance at the Alexandra Palace, a lavish building that covered seven acres and included a museum, lecture hall, library, banquet rooms, a 3,000-seat theater, a 3,500-seat concert hall, and a grand hall in which 12,000 could enjoy the thunderous sound of a pipe organ powered by two steam engines. Set amid a lavish 196 acres of parkland, it was a far cry from Rockaway Beach, and Baldwin must have been impressed as he walked the grounds with Farini.

But with such a high-profile venue came protest from lofty circles. Baldwin's planned jump so raised the ire of the authorities that the House of Lords found the topic worthy of debate. The Earl of Milltown wanted to know if the Home Office knew about this, and if so, "whether measures would be taken to prevent so dangerous and demoralizing an exhibition." The government said it would send the police to "warn the intended performer"[21]—as if Baldwin were a dumb Yankee who had no idea what he had gotten into.

Stoked by the publicity, forty thousand people turned up to watch the first show on July 28, 1888. By this time, Baldwin, with Farini's help, had improved the parachute's design by replacing the wheel at the apex with a circle of rope, making the chute lighter and entirely flexible. (If Farini hadn't helped with the design, he at least got his name on the patent as coinventor.) The jump went perfectly, but the press still railed. The *Court Circular* worried that next time Baldwin might be "dashed to pieces in the presence of thousands," and concluded that the stunt "can only appeal to the morbid interests of the mob, and should certainly be at once stopped."[22] Baldwin ignored the attacks and made ten more jumps at the Alexandra Palace, eventually drawing daily crowds of one hundred thousand. The Prince and Princess of Wales attended his final performance. He then went on tour, to northern England, Scotland, Australia, and the Far East.

Almost immediately, the simplicity of Baldwin's new parachute design inspired imitators and innovators. By the late 1890s, there were scores of parachutists dropping from the skies as far away as India and Hawaii. In Germany, Kätchen Paulus was among the first to stuff the parachute in a bag; she made 147 jumps before retiring in 1909.

In 1908, the strength of the new design was inadvertently tested when showjumpers Dolly Shepherd and Louie May—who would be jumping for the first time—were attached to parachutes and lifted above the English countryside. Clutching their trapeze bars, with three thousand feet of air beneath them, Dolly, by then a veteran jumper, told the new girl to pull the release cord. She did, but nothing happened. Terrified, she desperately kept trying as the balloon rose higher, with Dolly shouting encouragement. Nothing worked. Finally, at more than ten thousand feet, Dolly realized that eventually they would both lose their grip and be killed before the balloon descended on its own. Incredibly, Dolly let go with one hand, reached over, grabbed Louie, and told her to let go of her own trapeze and wrap herself around Dolly. She did. Dolly pulled her cord, they dropped, and the parachute opened—but only partially. Finally, the canopy filled, but they hit hard: Louie was uninjured, but Dolly broke her back. Nonetheless, one chute had saved them both—and Dolly went on to become the most popular British jumper of her time; she retired in 1911.

Around 1905, American Charles Broadwick was the first to pack a chute into a backpack attached to a body harness, which he used to scare the daylights out of his audience, since it appeared that he was jumping with no parachute at all. (His chute was released by a long static line attached to the balloon that was invisible from the ground.) Broadwick's design, made possible by the flexibility of Baldwin's design, would become the key to making the parachute a life-saving device.

Although Baldwin hadn't intended to, he had opened the door through which parachutists could move from the shadows of vulgarity back into the light of science. In later years, Baldwin himself would embody the change. He went on to design and build the first dirigibles in America. When Joseph Pulitzer offered the first Pulitzer Prize in 1908, a cash award of $10,000 for the first flight of any kind between New York City and Albany, two of Baldwin's airships competed against two airplanes piloted by pioneers Glenn Curtis and Wilbur Wright. (No one succeeded in making the flight; the prize was won the next year by Curtis.) When war came in 1917, Baldwin was appointed a captain, and later a major, in the U.S. Army, and was put in charge of inspecting the nation's fleet of dirigibles. Later he designed a legendary biplane called the "Red Devil," and formed a company to manufacturer the planes with Curtis. By the time he died in 1923, Major Thomas Scott Baldwin's name was writ large in the annals of aeronautics, and his beginnings as a wirewalker were hardly mentioned. Like parachuting itself, he had become respectable.

## Last of "the Greats"

In the twentieth century it was only in the circus that the nineteenth-century notion of the noble gravity daredevil lived on—or, more accurately, was carried on by habit. In reality, no performer managed to surpass Blondin in either skill or popularity, yet the tradition of pomp and grandeur, of battle-inspired drumrolls and dramatic cape-doffing, endured, in an effort to portray every trapeze artist or wirewalker as a hero. This was the tradition into which Karl Wallenda, who would turn out to be the last of the circus's "great" gravity performers worthy of the adjective, was born.

Wallenda was a throwback, a wirewalker whose audacity under the threat of falling commanded profound awe from an audience, as had performers a half-century before. Wallenda had a hunger for the wire, a lust for truly testing himself against gravity—and anyone could see it. "To be on the wire," he said, "is to live; everything else is just waiting."[23]

Born in a circus wagon in 1905 in Germany, Karl had five generations of circus in his blood. In the late 1700s, the Wallenda family was among those that had been transformed by the growing appetite for gravity acts and by the new venues, the circuses, that fed it. If you were a Wallenda, you started as young as two. Most couldn't remember a time when they hadn't performed. When he was ten, Karl was playing beer gardens and restaurants; at sixteen, he was doing handstands on the feet of a man doing handstands on a wire sixty feet above the floor of the Zeltsgarten vaudeville house.

By the time Karl reached his twenties, the future looked grim in a post–World War I Germany, where people were using wheelbarrows of money to buy groceries. The pay was better when they traveled, and after a performance in Cuba, when John Ringling offered to sign them to perform in his circus, Karl accepted and took his new wife and his brother Herman to America.

They were the Great Wallendas, and their opening in Madison Square Garden in 1928 brought a fifteen-minute standing ovation from cheering New Yorkers. For the next three decades, Karl pushed their act beyond what anyone thought possible. Real risk was the family's stock-in-trade, and individual Wallendas ran through the wirewalker's repertoire to deliver it, doing headstands, leaping, and riding bicycles on the wire. But this was not enough for Karl, and he soon began to move wire-walking in a direction it had never gone before: he put more people on the wire at the same time.

This meant more of a thrill for the audience because of the way Karl did it: he did not string them out, bouncing over each other in a line. Instead, he bound them together, so the fall of one would mean a fall for all. In the Pinwheel of Death, Karl and Herman were linked with a steel bar; on the bar their wives sat in opposite directions, and then, gripping the bar behind them, leaned back and spun.

Karl then began stacking the family into pyramids, where they stood on the linking bars between men. In 1947, he unveiled the Wallendas' signature act, one that is impossible to watch without feeling terrified for their safety. The trick begins with two men stepping onto the wire with a steel rod strapped between them at chest level. From above, a third man steps gingerly onto this rod, creating a three-man pyramid that inches out a few feet from the platform. They are followed by another three-man pyramid, the two top men also linked with a steel rod. On this rod someone steadies a chair, and onto the chair climbs one of the Wallenda women. She sits as the six men bear her out. Then, toward the middle, she draws her legs up to the chair's seat and unfolds herself to stand. The Seven-Man Pyramid, as Karl called it, moved over the wire like an enormous insect, balance poles waving like antenna, a thousand pounds of people and steel on a strand five-eighths of an inch thick.

The Wallendas performed the Seven-Man Pyramid night after night for years without an accident, until January 30, 1962, at the State Fair Coliseum in Detroit. As they climbed to the platform that night, the loudspeakers came alive with "The Washington Post March." As it echoed through the coliseum, the spotlight caught Dieter Schnepp, the leading bottom man and Karl's nephew, as he was first out on the wire. It was only his second pyramid performance.

The pyramid formed and moved smoothly forward. Suddenly Dieter was in trouble—something with his pole. He shouted, "I can't hold any longer,"[24] and the pole was gone, bouncing off the wire. Karl and his brother Herman were on the second tier. There was no net—there never was in a Wallenda act. Dieter buckled; the attached rod yanked Dick Faughman, Karl's son-in-law, down behind him; then Mario, Karl's son; then Gunther, Herman's son. Karl and Herman toppled from their stance on the rod, sending Jana, Dieter's sister, sliding from the chair and into space. Karl lurched forward and caught the wire. Jana hit him, slipped to the side, and Karl locked his legs around the sixteen-year-old girl. Herman grabbed the wire, and so did Gunther. Dieter, Dick, and Mario hit the ground.

Gunther was the first to climb down. Blood gushed from Dieter's mouth. Dick, on his back, one knee up, took shallow breaths; Mario lay curled on his side, gurgling, foaming at the mouth. Soon they were surrounded by clowns trying to reassure the audience, who sat stunned, nearly silent. Sirens began filling the air.

Karl's nephew and son-in-law would die that night, and his son, Mario, remained paralyzed from the waist down for the rest of his life. The others in the troupe went on to finish the season, the tragedy perversely drawing larger crowds even though the Seven Man Pyramid was missing from the act. A year later, Karl talked them into trying it again on a practice wire twelve feet above the ground. They fell. The wire slashed Gunther's face, knocked out teeth, and he quit. Still, Karl pushed the family, and they performed the pyramid a dozen more times until Herman decided he had had enough and left wirewalking for good. Then the others began to drift away.

By 1966, Karl was alone but could not quit. He knew he needed a new appeal, and he created it by reviving the tradition of Blondin, inventing what he called "skywalks." These were outdoor stunts, pitting himself, as Blondin had, directly against nature. What was different, and new, was something inadvertent, but something Karl could hardly have failed to realize. Here was not a young Blondin skipping across the rope on his way to greater glory. Here, instead, was the last of the original Great Wallendas, an old man, struggling out on the wire.

In 1970, at the age of sixty-five, Karl announced that he would walk across the Tallulah Gorge in Georgia, a two-thousand-foot gap 750 feet above a rock-filled canyon. A video made of that walk shows Karl standing on the edge in costume: black pants, white peasant shirt, a red sash. There is a microphone clipped to him. His adult grandson, Enrico, tells him not to try anything fancy or he won't have a trick for next time. Karl agrees and steps out. "God is with me now," he says.

Forty feet out, the red sash begins to flap. "It's a little windy now," he says. "More wind than I expected." One-quarter of the way out, he pauses. "My god is with me. I believe everything will be all right." Slowly, he squats and rests his balance pole across the wire. Slowly, he lowers his head to touch it. Then, he lifts his feet from the wire and raises them skyward, a headstand resting on the point where the wire and pole meet. He lets himself back down to the wire, travels a little farther, and does another headstand before reaching the other side.

The skywalk at Tallulah Gorge was a success, but Karl's advancing age meant that if he continued to challenge gravity, at some point he would lose. That day came on March 22, 1978, when Karl was seventy-three years old. He had volunteered to walk between two hotels in San Juan, Puerto Rico, for free, just for the publicity. A week before, the old man had fallen in practice, fracturing his neck, yet he had insisted on making this appearance. The film of this walk shows the cable stretched 120 feet above the street, just yards from the resort's beach. As Karl and his grandson, who assists him, stand at the edge of the wire, gusts of wind blow in from the ocean and ruffle their clothes.

Karl, thin and a little unsteady, steps out onto the cable facing the camera. This time, he is not in costume. His brown slacks, long-sleeved white shirt, and bald head make him look like a businessman; the only clue to his status as a performer is his sash, and the sash is flapping ominously in the rising breeze. He leaves the building's edge standing erect. A quarter of the way across the wire, he hunches over his balance bar. The wind blows up tufts of his hair and pastes his pant legs to his thin ankles. A few more steps, his eyes are down. His pole dips, nearly upends; the wire jerks left and right. Now he rights the pole, but the wire is lurching.

He bends over, brings the pole slowly below his knees, but the rocking cable pitches him forward. His right hand drops the pole and clutches the wire; his left hand grips the pole, far out to his left, obeying the wire-walker's ancient commandment to never drop the pole. His feet slip from the wire, but it seems he's going to catch himself by sitting on it.

But his seat slips off the wire, his right elbow bends, with his left hand still clutching the pole. That end goes down, the other swings up under his right armpit, and he finally drops the pole, grabbing the wire with both hands. He swings under the wire, legs up; he tries to clench the wire but the old man's hands are ripped away. He falls, back first, arms spread. He falls, sliding onto an angled support wire; his hands burn down the cable only briefly before he opens his arms. The camera swings wildly to hold the blurred man in frame, loses him, then frames a limp body dropping away, far below. It disappears behind the building's edge; people run from the site.

AFTER KARL'S DEATH, some of the remaining Wallendas carried on, sometimes coming together as a family act trying to recapture the past, sometimes alone. A year and a half after his grandfather's fall, Enrico Wallenda walked successfully between the same two hotels in Puerto Rico. Some of the Wallendas even revived the Seven-Man Pyramid in the late 1990s, and performed in Detroit at the site of the tragic fall. Another troupe of wirewalkers attempted to surpass the Wallendas; they succeeded in putting an eight-man pyramid on the wire, but with a net to catch them if they fell.

The high-wire challenge to falling that epitomized Karl's work, in terms of courage and ability against real risk, hasn't been surpassed and isn't likely to be. Wallenda was a nineteenth-century gravity hero adrift in the twentieth century. All around him the tradition of the specialists—the wirewalkers, the high divers—was either long dead or dying, only occasionally revived by performers such as Phillipe Petit, whose fame flickered briefly after he walked between the World Trade Center towers in 1974. By the time Karl Wallenda died in 1978, the old world that had admired his kind of gravity heroics was long gone. In its place was a youth culture that admired antiheroes and saw gravity not as a fearful threat but as a convenient plaything.

Today, the Wallendas who are still performing have trouble finding work. "We used to work 48 weeks a year with only one day off a week," Delilah Wallenda told a newspaper in 1997. "Last year, we worked six weeks. We can't live off that."[25]

PART II

# Rebels, Cults, and Primates

With movies came falling of a particularly hazardous kind: untried stunts dictated by demanding directors. This image, although a reconstruction, accurately depicts the leap that Charles "The Thrill-A-Minute Stunt King" Hutchison performed in *Double Adventure* (1921, Pathé), in which he broke both wrists.

CHAPTER 6

# Rise of the Gravity Bums

The movies—*that's* where you get hurt. —STUNTMAN EDDIE POLO, FORMER CIRCUS ACROBAT, TRAPEZE ARTIST, AND PARACHUTIST

THROUGHOUT THE TWENTIETH century, the old image of a gravity hero slowly faded as a new image rapidly replaced it. Prosperity during the Roaring Twenties would drive roller coasters to the height of their popularity. Around midcentury an exploding youth culture, drawn to sensation and rebellion, would find gravity to be the perfect toy: it delivered a thrilling rush, and parents disapproved of it. Teens, especially in America, would begin to create dozens of new ways to exploit the sensation of falling for fun, spawning recreational cults filled with adolescents who saw their chosen sport as the basis of their identity—a process that continues to this day. But before these changes, a new invention would create a radically new kind of professional daredevil and would make the image of a human risking death by gravity more common than it had ever been: the invention was motion pictures, and the new gravity hero was the stuntman.

TYPICAL OF HOW movie stunts began is a story, perhaps only a myth, about the 1915 epic *Intolerance.* While shooting one day on his mammoth set, D. W. Griffith found he had a problem. For a key sequence of the movie, the storming of Babylon, he had re-created the gates of the city, enormous doors in a massive wall thick enough to drive a chariot along its top. He had hired hundreds of extras; crowds of ancient soldiers swirled beneath the battlements amid catapults and rolling assault towers; dozens more lined the top of the wall, hurling stones and spears. What he needed now was action of unquestionable realism, something stark and compelling to show what battle was really like. It wasn't enough that some of the extras had been doused with flaming oil raining from the parapets, or that he had a good shot of an actor being run through with a spear; it was still clearly an illusion. What he needed was real, on-screen danger. What he wanted was for soldiers to fall from the walls.

Normally, it wasn't hard to get extras to risk killing themselves on stunts like this. The principle was practically the cornerstone of movies at the time. Movies were about action not actors. Five years earlier, the wildly successful Keystone Kops series was created when Mack Sennett hired a crew of failed circus acrobats, ex-fighters, and various other roustabouts who were pliable enough to do what he asked. While tumbling or boxing ability came in handy, skill was less important than nerve. Sennett had hired an accomplished racing driver, Del Lord, to pilot the Kops' careening vehicle, but he also gave him assignments any fearless idiot could do, like roaring off the end of the Venice pier at forty miles an hour.

Unlike Sennett, though, Griffith didn't need people with skill. Stunting was not thought of as a skill at the time. There were no stuntmen per se—no elaborate staging, no training, no safety devices to speak of. Griffith simply expected one of the extras to jump from Babylon's seven-story wall into a hay wagon.

The day before the stunt, when the "brodie" had been announced (movie falls were called brodies, after Steven Brodie, the man who claimed to have jumped from the Brooklyn Bridge) with the promise of a three-dollar bonus added to the normal day-rate of five bucks, an old Native American everyone called Eagle Eye had volunteered. Eagle Eye, so the story goes, performed the stunt in his usual state: dead drunk. He dropped from the wall into the wagon, and everything was fine.

The following day, Griffith wanted the stunt performed again, but Eagle Eye refused. He had, he said, gotten religion and sworn off booze, and there was no way he would make another jump sober. But Griffith needed the shot. So he hired a minister who told Eagle Eye that just for today it was okay to get plastered to fulfill his obligations. Roaring drunk again, the old Indian plummeted into the cart.

In the early days of movie stunts, car crashes, explosions, and fights with wild animals were always part of the mix, but it was the peril of gravity that really delivered excitement, because these tricks weren't faked. For car crashes, directors would undercrank the camera, making the car appear to be going faster; for explosions, they would stop the camera just before the blast, move the actors out of the way, and then start again, making it look as if the actor had been in the middle of it. "Wild" animals were typically drugged. But gravity stunts were real.

Movie makers soon discovered that they could exploit the public's lust for gravity thrills by making films that were little more than dangerous stunts strung together by the thinnest of plots; these were especially profitable when produced as serials. In 1913 there was *The Adventures of Kathlyn*, featuring Kathlyn "The Girl Without Fear" Williams; in 1914, *The Perils of Pauline*, appeared; so did *The Hazards of Helen*, which eventually became a 119-episode phenomenon. It was so common for directors to leave their stars dangling from a precipice in an episode's last scene (the better to bring the audience back) that serials became known as cliffhangers.

From these beginnings well into the 1930s, stunt films filled with all kinds of harrowing leaps and falls marched across the screen, their titles alone telling the story: *Perils of the Yukon*, *The Grim Game*, *Taking Chances*, *Laughing at Danger*, *Look Out Below*, *High and Dizzy*, and *Safety Last*, which features the famous scene of Harold Lloyd hanging from the hands of a clock high on an office building.

Although falling to entertain wasn't new, there was an important difference between the gravity stunts now being done in the movies and those of earlier performers. For as long as people had risked their lives falling for an audience, there had always been a principle so obvious it hardly needed to be acknowledged, and that was that a gravity daredevil chose his own stunts. Sam Patch, Jules Léotard, the Great Blondin, and Karl Wallenda might have felt pressure to come up with new thrills, but they decided for themselves how much risk to take. The movies changed this. Now there were directors who envisioned what would best shock the public and who ordered stunters to comply or get fired. If the first hire got hurt or killed, there was always someone to replace him. Richard Talmadge, for instance, crossed a cable hand-over-hand sixty feet above a street just after the previous stuntman had fallen (and wound up crippled). In this case, the first man was his brother.

And in the movies, because many stunts were untried, their consequences were completely unknown. Gravity daredevils of the past had done essentially the same stunt over and over, pushing the envelope if they could. A wirewalker, trapeze flyer, or high diver had history to draw upon, both from others who had gone before and from his own experience of perfecting his skill. But who knew what would happen to someone riding a horse off a fifty-foot cliff, a stunt performed for *The Hazards of Helen* serial (both lived) or a jump from a moving train into a lake (the stunt man missed the lake and was killed).

Even when there was plenty of knowledge about a stunt and about what would keep a performer safe, in the movies this was ignored. For someone risking a fall, nets were almost never used. Safety consisted of a hay wagon, a pile of mattresses, or just a little extra dirt. In *The Perils of Pauline*, the star's double worked his way across a high wire protected by one man holding a carpet tied to a picket fence.

Typical of the breed willing to face risks like these was Charles Hutchison, or Hutch, as he was known, a stage actor who saw the movies as a new opportunity for employment. Hutch soon found that risking his neck brought more jobs than straight acting.

In 1918, Hutch and his fiancée, Edith Thornton, were hired by Pathé for several serials that revolved around stunts. Hutch carried Edith across a log over a canyon in one; in another, they flew off an embankment on a motorcycle together—the crash temporarily paralyzing half of Edith's face. In one of Hutch's solo stunts, he dove from the roof of a two-story house to a tree twenty feet away and broke both his wrists.

Pathé began to bill Hutch as the Thrill-a-Minute Stunt King, and the pressure was on for the studio's directors to deliver the goods. Sometimes they would think up stunts on the spot. One director demanded that Hutch jump from a balcony to a chandelier and swing across the room. No one bothered to see if the chandelier would hold. It didn't. Hutch crashed to the floor covered with glass, and snapped his elbow out of its socket so it bent the wrong way.

By 1921, fear had caught up with Hutch. He began to refuse to perform the most dangerous feats, and others would have to be brought in to double him. By 1926 the studio tired of a stunt king who wouldn't do stunts, and fired him. Hutch's career was over.

But there were plenty to replace him. In the Los Angeles of the early 1900s, a rootless city of people from somewhere else, it wasn't hard to find foolhardy men. Besides actors like Hutch who made the switch to stunting, there were refugees from the circus, rodeo riders who had drifted in from vanishing ranches, barnstormers who had come to perform for crowds above the amusement piers along Venice Beach, and an assortment of itinerants who had simply gone west until they had run out of land. Sennett described one of his men this way: "Hank Mann was the toughest. He stayed with me for eight years, and worked up to $135 a week, for which he would perform any foolish stunt my psychopaths in the writing room could think up."[1]

The movie men soon discovered that falling could be made more exciting if either the thing you were jumping from or to—or both—was moving. This made split-second timing critical, and a miss, even at relatively small heights, lethal. In one scene, Johnny Stevenson jumped from a bus to a bridge girder, slipped, and was killed when he hit the street, eighteen feet below him.

Moving jumps became appallingly dangerous in short order. In 1919, Ormer Locklear climbed down a ladder hanging from a speeding plane and dropped into the cockpit of a plane just below. It was a stunt worth repeating, and he was hired to do it again in 1920 for *The Skywayman.* But before he could complete it, Locklear died in a plane crash stunt for the film. Ted McLaughlin replaced him in the plane-to-plane transfer shot; he was killed when his ladder swung into the propeller.

In 1925, Gene Perkins tried a trick that would have been spectacular if it had worked. The plan was for him to drop from the end of a rope ladder suspended from a plane to the top of a speeding train. With Perkins holding on to the bottom rung, his pilot swooped down toward the train but missed the approach, and so circled around to try again. He missed again on the second try, with Perkins still hanging by his hands. On the third try he came in too low. Perkins slammed into the side of a railcar but hung on as the plane veered off. Finally, he lost his grip. He hit the ground, fifty feet below, at perhaps a hundred miles an hour: the impact drove his leg bones through his shoes; his massive internal injuries killed him in a few hours.

In 1927 Richard Grace decided to try Locklear's plane-to-plane transfer, but with a twist: he wouldn't use a ladder. In other words, he would simply drop from one plane to another in midflight. In his biography he describes what happened as he began the stunt, hanging by his knees from one of the higher airplane's wing struts:

> Unlocking my knees, I let my body slide down toward the earth. . . . Then, in a split second, while falling through the air, I realized that I had misjudged. It might have been because of the whipping clothes that bothered and hampered me in leaving, or it might have been that the rush of blood to my eyes had affected my judgement, but from whatever cause I realized that I would fall in front of Tomick's ship—

> not by mere inches but by at least a foot and a half. . . . Again there was the necessity of instantaneous decisions, and my decision was not to try and divert my body from its course but to try and catch a cabane [a strut wire] as I passed the leading edge of the top wing of Tomick's plane. I swung my right hand backward, and made the one attempt to grasp the thin small wire. For a long second I thought I had missed. Then the two forefingers touched and slid over it—and closed. My body turned a complete somersault. Still those fingers held. With a sudden jolt my feet hit the second bay strut and I hung in front of the plane.[2]



With tricks like these becoming a movie staple, stunt people began dying at an alarming rate, and the press rushed to cover the most spectacular deaths. Three men died during the 1929 filming of *Hell's Angels*, under the direction of a demanding Howard Hughes, and headlines blared the fact.

In the five years between 1925 and 1930, fifty-five people were killed making movies, and more than ten thousand injured. Now under scrutiny, experienced stunt men began to separate themselves from amateurs by building special equipment, rehearsing stunts, and developing new techniques. On a stagecoach drive over a cliff for the 1940 film *Dark Command*, Yakima Canutt fashioned an elaborate cable system that would yank back the plummeting coach before it fell on the stuntman and horses; he also created a breakaway harness from which the horses were released before they ever hit the water.

By the late 1930s, the maverick stuntman willing to do anything for a buck was disappearing. In 1937, the Screen Actors Guild, which included stunt performers, was accepted by producers; it established minimum pay as $25 per day, $35 for stunts, and $5.50 for extras. Safety improved dramatically, and just as with circus performers, knowledge began to be shared and to be passed from father to son, and eventually from mother to daughter, as entire families began to work in the business.

With the rise of professionalism, stunt performers became anonymous. Unlike the early stars, their names were unknown to the general public, and their stunts safe enough so that no one went to the movies expecting to see anyone actually risk their life.

## Do Burns, Stair Falls, Wrecks—but Don't Do High Falls

While all movie stunts today are far safer than in the past, high falls remain the most dangerous, and so are performed by specialists. There are only a few-dozen men and women in Hollywood, and probably fewer than a hundred worldwide, who will fall farther than forty feet for a movie shoot; fewer still who will do a flip or spin as they fall. If you need a woman to drop from, say, eighty feet and turn an out-of-control back flip on the way down, you're looking at perhaps two or three candidates. One of these is Nancy Thurston.

Nancy's a blonde with a winning smile and a petite frame that tends to disguise her athleticism. By her intelligence, you might mistake her for a young business graduate fresh from an Ivy League school—until you see her video. In it, she turns handsprings across a field, executes a perfect two-and-a-half somersaulting high dive from a three-meter springboard, sails from a flying trapeze, backflips into a fight to knock down a pair of brawny bad guys, gets hit by a car (which flips her over the roof to slam into the asphalt with bone-crushing impact), and leaps from the top of a ninety-two-foot cliff across a mountain of boulders.

Nancy performs the kind of stunts her friends in the business have been warned not to try. Nancy tells me that one of her friends, whose dad was also a stuntman, was cautioned by his father. "I don't care what you do in this business," his dad had said, according to Nancy. "You can do burns, you can do stair falls, you can do wrecks. But don't do car hits or high falls."[3] The reason is simple: in spite of greatly improved technology, these are still the stunts that most frequently hurt stunt people, and of the two, high falls are the worst: a mistake can kill or paralyze. In the nine years Nancy has been in the business, she's seen two stuntwomen friends die and two others become quadriplegics when a fall went wrong. In each case, there was nothing wrong with the precautions taken—the accidents were solely the performer's fault. And even if something was wrong with the planning, in the stunt world, it's considered the stunt person's responsibility to make sure it's right.

I learned this from Jeannie Epper, Nancy's friend, and president of the Stuntwomen's Association of Motion Pictures, for which Nancy is a group representative. Like many people in the industry, Jeannie was born and raised in the stunt world; her father did horse stunts in the 1930s and 1940s, and at present, fourteen of his descendents, including his great-grandchildren, are stunt people.

On one particular shoot, Jeannie was to flip over a guardrail, protected only by four men holding a cable. "I was falling backwards eighteen feet to the ground," she told me. "I'm not going to *hope* these four guys don't trip and fall and drop me on my back." Jeannie stopped production and called in her son Richard, an expert rigger, which caused a showdown with the coordinator in charge of the stunt. Jeannie invited the coordinator to try it, and when he did, the rigging broke, dropping him to the ground. "It didn't hurt him very badly," she said, "but it made a bigger ass of him, and now he was really furious." Her son then stepped in to suggest that one safety line be added, and the crew got their shot.

In a shoot that involved her brother Gary, things were more serious. Gary was set to hang over a cliff on a motorcycle with stuntwoman Sandra Gimpel—but Gimpel questioned whether the eyebolt screwed into the earth would hold. She too had to stop production and insist that the cable also be secured to a nearby tree. When the camera rolled, the eyebolt was jerked from the ground—they were stopped from falling only by the second anchor that Gimpel had demanded.

In high falls, such as the ones Nancy does, safety isn't usually a question of proper equipment but about the circumstances of the fall itself. Two years ago she had to turn down a 103-foot fall that she had looked forward to, once she saw how it was set up. "I had wanted to go around a hundred; my highest is ninety-two, and I consistently go eighty. It's just my own personal thing." The script called for her to do a flailing backflip from a ten-story building. She knew the height and site beforehand, but when she arrived she discovered that because of a furnace abutting the building, her air bag had to be positioned fourteen feet out. "When you're up there looking down, it looks like you can't miss the air bag, it's so big. But that's why we measure it out." The stunt would have required that she push off hard to get the distance, something she would have had to carefully rehearse. On the advice of a trusted mentor, Jon Epstein, she turned the stunt down, and a stuntman was brought in to do it. "He pushed like he went off an air ram," she told me, "and he *just* landed in the middle"—that is, he barely made it to the part of the bag that would absorb his fall rather than dangerously bouncing him off to the side. It was hard for Nancy to bow out. "It was an ego thing for me. I had friends who had flown up to San Francisco with me to watch it. But I had to think about what Jon had told me. I have to say Jon saved my life that day."

Jeannie, too, has declined stunts. For a job in the feature film *Romancing the Stone*, she was to leap from a car that was tumbling over an eighty-five-foot waterfall. "If you don't leave the car before it falls, you can't project yourself away from it," she explains. "You have to leave something that's steady. You can't leave a falling object, because you'll stay right next to it." Unsure that she could leave in time, she let stuntman Terry Leonard fill in—and he nearly found trouble. "Terry didn't get pushed off soon enough. He left just as the car started to tip over. Whether he slipped, or whatever, he didn't quite get off enough, and he almost got killed. He got sucked underneath the water, and couldn't get out from under the waterfall."

Calculating risk, and being brave enough to know when it's too high, is a part of the business the two women accept. "I don't care how long you do high work," Jeannie said, "it's an unnatural thing. Your brain always says to you, 'Why are you stepping off this building?' Dar Robinson used to tell me—and he was a guy who broke all records—'unless you can sit down and draw it out on paper, I'm not going to do it. I have to know that the physics and everything are right, because my brain says to me, you can't do this. You're going to die.'"

What protects Nancy is the skill she's acquired over twenty years of springboard diving, beginning when she was five. By eight, she had decided that being a stuntwoman was what she wanted to do. After some time spent in high-diving shows at Magic Mountain and other amusement parks, she entered the profession with the help of Bob Brown, one of the leading high fallers in Hollywood.

Falls in the movies begin with a stunt that's been written into a script and then planned and budgeted by a stunt coordinator. When the job is offered to a stunt person, there's usually some negotiation on both sides. A fall can bring anywhere from $200 to $2,000 dollars or more, depending on the budget of the movie, the height, the circumstances, and any special requirements, "like if you're on fire," as Nancy says. This is in addition to union scale for a day's work, which is usually around $500. Sometimes a stunt person might suggest a lower fall if the one planned doesn't seem safe, or might offer a higher or more elaborate fall, such as a flip, that the coordinator might not have thought of and that will bring more money. Occasionally, there will be an argument on the day of the shoot. Jeannie won one of these when she was doubling for Wonder Woman in the old television show. "I was jumping off an old, dilapidated building's fire escape—just the platform, with nothing around it," she told me. The director hadn't wanted to pay what she thought was a fair price, so she invited him up. "And he gets up there, and you could see the color drain out of his face." She got her price.

The hay wagons that used to catch high fallers in the days of D. W. Griffith have long been replaced, first by porta-pits, which were large canvas bags filled with chunks of foam rubber, and then by cardboard boxes that would be stacked and covered with a tarp. Surprisingly, boxes worked well, and Jeannie remembers the security she felt from them when she was doing an unusual kind of fall for the *Wonder Woman* show called a reverse shot. In this, she would stand at the top of a thirty-foot building facing backward, with her arms raised. She would then leap off, her head looking up, her body erect; she would hit the boxes in a standing position. She told me she liked the firmness they offered before they were crushed. The reason for the odd posture was that the film would be run backward, making it appear that Wonder Woman had just leaped to the top of a tall building in a single bound.

Today, boxes for the most part have been replaced by air bags. These look like enormous rectangular air mattresses six feet thick. Inside, there are two chambers; the lower one is inflated until it's firm; the upper one less so. The top chamber has flaps that cover large vents along the side; on impact, the vents fly open, air is released, and the faller gets a relatively soft ride to the surface of the firm bottom chamber.

The most recent advance in the technology of falling are decenders and decelerators; both are cable systems that hook to a harness worn by the faller. Decenders control the fall all the way, while decelerators break a fall at the last moment.

None of these systems is perfect. The cable systems offer the best protection, but they can't be used if the camera will see the cable, or in falls in which a stunt person turns or flips. Air bags are the most widely used protection, but to be effective, you have to hit them within a safety rectangle marked on the surface. If you hit off to the side, the bag acts as a catapult. This is what happens to most high fallers who are killed or injured; they're thrown from the bag into a wall or pavement. The bags themselves are nearly 100 percent reliable, but in one case, on the set of a movie called *Steel*, one failed. The stuntman who died did everything right; he was an expert at high falls and plummeted an incredible 320 feet, landing in the center of the bag. "When he hit," as Jeannie heard the story, "it just opened up and dropped him right to the concrete. Everybody thought he had landed just fine. Then he doesn't come out, and he doesn't come out." It was discovered later that the bag's seams had been damaged by temperature changes when it was shipped by air. In spite of the tragedy, the scene was kept in the movie's final cut.

As skilled as today's stunt people are, they still sometimes have difficulty earning the respect due them, and in part this is because of the tradition of daredevilry born in the early days. "I don't know if we've ever truly, honestly trusted the fact that we're smart," Jeannie told me. "Because we're so physical, people mistake that for lack of intelligence rather than understanding that what we do is precisely calculated." The business is especially hard for women. "We have to go in with a lot more knowledge," she said. "It doesn't mean that the guys don't have as much knowledge, but we're afraid if we only have as much as they have, we won't be accepted. So we have to have a little more. Let's say I hire out to do a car stunt, and I overshoot my mark and hit a camera. 'Shouldn't have hired her.' I've heard them say that. But if a guy does it, it's just, 'Oh well, he's having a bad day.' This is just the way it is in Hollywood." Women also have to consider costumes in a way men don't. Nancy mimics coming to a shoot: "Here's your wardrobe," she says, holding up pinched fingers as if there's a bikini dangling from them. "And I'm crashing onto the ground in *this*?"

Added to this is a new threat to their livelihood: computers. Now that people can be created digitally, dangerous—indeed, impossible—stunts can be put on screen with no more than a risk of carpal tunnel syndrome. A digitally enhanced car hit, for example, opens the film *Meet Joe Black*, a stunt that would have obviously killed anyone trying to perform it.

But in the near future, the elimination of professional fallers isn't likely; they will simply help digital animators by falling in more controlled situations, with the dangerous circumstances added later. "It could save some lives," Jeannie notes.

In the meantime, Nancy, whose career is just beginning, continues to enjoy it, especially the very high falls. "Sometimes it's like slow motion," she says. "You feel like you're in the air forever. That's why I like it. You're just suspended in the air. It's an incredible feeling. Some things you can't really explain. You just have to experience it."

## Free-Falling

When airplanes began to appear in the early part of the twentieth century, it would seem that strapping parachutes to pilots would be an obvious and simple way to protect them. By then parachutes had been made flexible by Thomas Baldwin, put in a pack by Kätchen Paulus, and fixed to a body harness by Charles Broadwick. All that remained was to put the cord that released the parachute in the hands of the jumper. But there were problems.

At first, planes flew too low for a parachute to work; in 1909, the altitude record was 505 feet, and a typical flight was lower. As they began to fly higher, a parachute made sense—but the planes were also flying faster, and the question became, how could a pilot flying at a hundred miles an hour get out of a cockpit and have a parachute open without it getting tangled up with the plane. He could certainly jump out with a parachute on his back and pull a cord that would release it, but everyone knew this would never work.

It wouldn't work because that would mean the pilot would fall freely through the air first, and it was believed that this alone would kill him. Experts agreed that the speed of a free fall would knock him unconscious by a hundred feet and then suck the air from his lungs to suffocate him at five hundred. At the very least, the wind would pin his arms back, making it impossible to release the chute.

Two men knew this was nonsense, and for the same reason. In addition to being pioneering pilots and show parachutists, Leslie Irvin and Floyd Smith had spent time on the flying trapeze; Irvin had tried it a few times for fun, while Smith had been a professional, one of "The Flying Sylvesters." Both had fallen from nearly a hundred feet into nets without ever feeling faint or having the slightest trouble controlling their bodies. And Irvin had another reason for his belief, drawn from a careful, if gruesome, observation. Irvin had once been standing on a tarmac within yards of a man who had jumped from a plane but whose parachute had failed—and since the man was screaming and flailing up to the moment he hit, Irvin concluded that he could breathe and move.

In September of 1918, Smith was appointed by U.S. Army General "Billy" Mitchell to head a team commissioned to solve the parachute problem. During the war, Europeans had made headway with chutes mounted on planes that would open automatically (to avoid the dreaded free fall), but these were complicated, failure-prone packs loaded with springs, compressed air, even gunpowder to blast the chute skyward. In one design, the pilot was to wear a special hat. When it came time to bail out, he was to throw his hat overboard; its spring-loaded wings would then pop out, catch the wind, pull the main chute out, and pluck the pilot from his craft. The design that finally worked reasonably well was a static line attached to the plane and to the stowed parachute. When the pilot climbed out and jumped, the line would release the chute, although this plan was still unreliable because the line would sometimes snag and break on the fuselage. Pilots were naturally reluctant to use them, and Mitchell was determined to find something better.

Smith came to the team with something he believed *was* better. Only three months earlier, he had patented a design for a body-pack parachute that for the first time included the release cord on the pack itself. But using it would require free-falling. Indeed, since the point was to pull the rip cord during the fall, even testing it would require a human to free-fall—and no one was ready for that yet.

So Smith diligently tested dozens of designs that relied on a static cord, including ones made by Irvin. They would fling a rubber dummy from the cockpit during spins, stalls, and dives, to see what happened under mock battle conditions. Things didn't go well: the dummy hit the ground so often beneath a failed chute that they named it "Whistling Jim."

But it gave Smith the evidence he needed. He concluded that "the operating means must not depend on the aviator falling from the aircraft."[4] In other words, fuselage-mounted parachutes released with static lines could never be made reliable. That being the case, Smith's own design would have to be tested. Someone would have to free-fall. He asked Irvin, the former showjumper who seemed fearless: as "Sky-Hi" Irvin, part of his act had been to sit on a parachute's trapeze and then "slip" off backward, to hang by one foot. Since Irvin also believed in the rip-cord idea, he agreed.

On a spring day in 1919, Smith's team, along with Irvin, gathered at McCook airfield. Smith took the pilot's seat. As the plane climbed into the cooling air above Ohio, Irvin began to have doubts for the first time. "I knew I could do it," he said later, "but everybody else was so certain I couldn't, I began to doubt my own confidence."[5] At fifteen hundred feet, Smith leveled the plane, and it was time. Irvin stood up in the cockpit and then dove headfirst over the side. The instant he felt he had cleared the plane, he yanked the rip cord with both hands. The chute opened perfectly.

It wasn't much of a free fall, but it had been done, and Irvin had survived. After a few more tests of static-line chutes, including some that killed their jumpers, the army, and a few showjumpers, carefully adopted the new design.

But doubts lingered. Irvin had opened his chute immediately; what if a pilot had trouble finding the rip cord? How far could he fall before he was rendered unconscious—a consequence of free falling that most still believed inevitable? In 1924, Sergeant Randall Bose attempted to find out, and his report was alarming. After jumping and then waiting only thirteen seconds before pulling his cord, he reported that "things started to go black before my eyes. I was spinning very violently, and feared I might lose consciousness."[6]

The experts were worried. But, as usual, they were ignoring the experience of showjumpers who were already pushing the limits of free falling. In 1922, barnstormer Wes May would roller-skate on a biplane and then "slip" over the edge into a free fall, horrifying the audience for as long as he could before pulling the rip cord. The following year, Aaron Kranz and Art Starnes were also free falling—"bullet drops," as they were advertised. In 1925, Joe Crane set the record by falling thirty-five hundred feet before opening his chute.

In the years that followed, performers took up free falling with relish, using it to pack more falling into a single jump. Buddy Plunkett, who started jumping at age fourteen, would fall from a plane and release a torn parachute that would stream above him like a hideous limp shroud while he fell several thousand feet. Then he would release his reserve.

In the 1920s, Jack Clapp would ride up in a balloon, jump out, let a parachute out of his pack, float down for a while, cut it away and plummet for another thousand feet or so, let out another parachute, cut that one away, fall again, and finally let loose his third parachute just in time to drift down the last few hundred feet. Clapp knew why people liked this trick. As he put it, "They figured a fellow using three parachutes had three times as many chances of breaking his neck as a fellow who used only one."[7] Clapp stopped jumping at age seventy-five; many others who tried the cutaway trick had their careers come to an abrupt end when they opened their last chute too close to the ground.

Although they probably didn't know it, these men were the first in history to experience the ultimate fall: that is, after falling fifteen hundred feet, they were falling as fast as was possible, traveling at 120 miles per hour, what would later be called terminal velocity. What they did know was that free falling, while feasible, was far from easy. The spinning that Sergeant Bose had experienced had hit them as well. Again, a showjumper discovered the solution. On March 1, 1931, Spud Manning jumped out of a plane more than 16,000 feet over Los Angeles and let himself fall to within fifteen hundred feet of the ground. He had fallen three *miles.* Manning was tight-lipped about how he did it—that was a trade secret for a showjumper—but Floyd Smith knew, and after Manning's death in 1933 (his plane crashed into Lake Michigan), Smith wrote an article describing Manning's maneuvers: his basic stable position, as well as his turns, dives, and spins. Although free falling—its techniques later formalized and named "skydiving"—wouldn't be developed as a sport until decades later,

Manning had discovered not only that you could fall as far as you liked without being suffocated, but also that the experience of falling had a new appeal that had been hidden for all these centuries: if you fell far enough, it felt like flying.

## Deep and Thrilling Dips and Plenty of Them

The sensation—and sensationalism—of falling rolled across America in continuing waves throughout the 1920s: showjumpers "bullet dropping" at county fairs, stuntmen leaping off cliffs in the movies, and—in a surprisingly sudden burst of interest in a gravity device nearly as old as the parachute—roller-coaster mania.

After its invention in 1784 in St. Petersburg, several coasters had been built in Russia, France, and England, but they only amounted to a handful. America had none. This young country founded on rebellion apparently hadn't needed simulated heroics; its people had plenty of thrills just dealing with a wild new world. But by the late 1800s, the frontier was closing, and probably not coincidentally, the American roller coaster was born.

It was inspired by the nation's first commercial railroad. Originally built in 1827, the Mauch Chunck Railway was designed to bring coal nine miles down Summit Hill in Pennsylvania to the Mauch Chunck landing on the Lehigh River. Since steam locomotives didn't exist then, the descent was powered entirely by gravity, the cars being hauled back up by mules. By 1872, a new route using a steam locomotive was built, and the old railway was converted to carry passengers in open cars as a tourist attraction. In a nod to the popularity of all things Alpine, it was advertised as taking riders through "The Switzerland of America."[8]

The Mauch Chunck Railway bore little resemblance to the short, straight-down slopes of European coasters—for one thing, the round-trip was eighteen miles—but it produced excitement similar to that of modern roller coasters, using the same techniques: slow, steep ascents that create anticipation, followed by exhilarating downhill runs complete with bumps, sharp turns, and fast runouts.

As a rider, you would begin by taking a seat on one of the several wooden benches in the open-air car. Directly ahead was a steep incline "hung as it were like frail bits of thread between the heavens and earth," as one rider put it, adding, "Are we to go up that lofty ascent, steep as the roof of a house, and ending among the clouds, and in a car too?"[9] Behind your car a *barney* (a small car designed solely to push the larger car) would appear. You would hear a bell ring, and the ride would begin its climb up the back road to the Summit Hill.

The barney would start to push, and the smells of the river would give way to those of the pine forest surrounding you on either side of the raised track. The barney was drawn to the top of Mt. Pisgah by a long chain of metal plates that ran from the bottom of the barney to a winding mechanism powered by steam at the mountaintop—an arrangement very similar to the modern coaster's chain lift, and apparently just as frightening. "Good heavens!" commented a tourist, "how we seem to swing in the air! If anything holds us to the firm earth or the heavens above, and keeps us going up instead of down 'ker smash,' it is these frail little plates trundling nimbly upward on rollers. What if one of them should break, which it is liable to do, and then how quickly the other would go!"[10]

As you approached the summit, you would hear the grinding of the winding mechanism growing louder. The barney would give a final nudge and send the train down a gentle slope on the other side. Passing over a trestle, you would come to stop at the Pavilion Station, where you could disembark and picnic. Back on the train, you would now face another incline up Mt. Jefferson, and another gentle ride down, to the Summit Hill station. Once again, you could hop off the train if you like.

When the ride resumed, your train would simply be released at the top of Summit Hill. The slope was so steep that the railway designers had to build switchbacks, producing a series of hairpin turns down the mountain. At the Mt. Jefferson Passover, a bridge where you would cross above the back road heading up, you could wave at the passengers below as you sped by. Parts of the road here were old and uneven and would bump you up and down in your seat, as you spun through the turns and accelerated down the straightaways.

Then you would hit the Home Stretch, a two-and-a-half mile straight track where your train would reach the speed of fifty miles an hour—and where the bumping was most severe. Another stop at Lentz Trail Crossing, and then down again, until the brakes were applied and you rolled slowly beneath the roof of the Mauch Chunck Depot.

Tame as the ride might seem compared to today's coasters, it was a mighty dose of gravity for people who had never felt such a thing, and they were enraptured by it. "Faster and faster," enthused one T. L. Mumford, "down through long stretches of shaded roadway, around wondrous curves, along giddy cliffs, under shadows of great ivy-grown crags, and still down, down, down, at a dizzy speed, and as if borne on the wings of the wind."[11] The railway was so popular that special excursion trains streamed in from New York and New Jersey, helping to swell the crowd of passengers to thirty-five thousand annually.

In 1884, clearly influenced by the success of the Mauch Chunck "Switchback" Railway, La Marcus Adna Thompson, a former knitting factory manager, built what he called his Gravity Pleasure Switchback Railway at Coney Island. It wasn't intended to be a thrill ride; Thompson advertised it as a scenic tour along the beach. As such, its cars rolled down a six-hundred-foot undulating ramp at a leisurely six miles an hour. The ride was an instant hit, attracting up to twelve thousand people a day. Thompson expanded his business rapidly; only four years later he had built fifty rides, and like an early version of Disneyland's Matterhorn ride, he often had his tracks run alongside elaborate scenic dioramas and through caves and tunnels, all enclosed in a gigantic faux mountain.

In spite of Thompson's success, the scenic coaster didn't last long, as other designers focused on delivering a bigger kick. A year after the Gravity Pleasure Switchback opened, Phillip Hinkle invented the chain lift, allowing a coaster to be cranked up a hill of practically any size. But it still couldn't be any steeper, because the car would fly off the track. This problem was solved in 1912 by John Miller, a coaster designer who had started with La Marcus Adna Thompson. His patented wheels attached under the track and kept a car on course even if it was sent straight down. With these two inventions, American coaster designers pushed the sensation of falling far beyond what riders had felt on the Gravity Pleasure Switchback Railway—indeed, far beyond anything that had been built in Europe.

According to Miller, this was no less than what people demanded. "The amusement-seeking public of the present day desire more than a ride when patronizing the modern amusement park," his 1923 catalog read. "They want speed, sensationally deep and thrilling dips and plenty of them, and Mr. Miller has solved the problem by catering to these most exacting desires in his new type of construction and cars."[12] Yet Miller also knew the implications of his invention, adding matter-of-factly, "More caution is required in designing the new type of construction in order to keep within the limit of centrifugal gravity, obtaining the sensational and thrilling dips demanded, yet not reaching that extreme which would pitch the passengers from the cars."[13]

This worry was hardly unreasonable for coaster builders of the time. In San Diego, at the Giant Dipper that rises on the beach at Belmont Shore, one of the original cars from this 1925 roller coaster is on display. It features two benches that are about as braced against gravity as those in a golf cart. The car has no lap bars or seat belts, and the sides are open. It looks like a breakfast nook on wheels, not something designed to keep people from being hurled over the side during the car's fifty-five-mile-an-hour plunge down its first hill.

Unfortunately, pitching people out of a coaster's cars was something that occasionally happened, along with other problems, and Miller himself was one of the contributing culprits. Over the years he designed dozens of frightening coasters, and one, the Puritas Springs Cyclone, acquired an especially fearsome reputation, even though it was built decades after Miller had issued his warning. It flung one man out on a curve; he fell forty feet to the ground and nearly died. The Cyclone's first drop—or rather the sharp uplift after the drop—had a tendency to jam people's vertebrae together. In the summer of 1942, three women were hospitalized with fractured spines, and the Cleveland police shut the Cyclone down. The ride's owners agreed to slow the speed on the drop but insisted that the Cyclone was safe as it stood. "The women that were hurt had weak backs,"[14] a spokesman stated flatly.

These blips in the safety record, though, didn't stop America from becoming the undisputed center of the roller-coaster mania of the 1920s. Along the way, roller coasters created a new institution: the amusement park, a place filled with machines designed to deliver a variety of physical sensations, most of which would exploit the sensation of falling as their primary appeal. Before they were transformed by gravity thrills, these parks had simply been a collection of picnic areas, restaurants, and dance halls located at the end of rail lines. They had been created after the Civil War by builders of the newly urbanizing cities' trolleys who created the attractions to increase ridership on the weekends, when commuters typically stayed home. Later in the century, entrepreneurs took note as roller coasters began to show themselves to be extremely popular. In the early 1890s, Captain Paul Boyton created the first two parks to feature mostly gravity rides, the Water Chutes park in Chicago and a similar park on Coney Island. This inspired the trolley men to add roller coasters to their parks, and their popularity led to parks going up everywhere. By 1919, more than fifteen hundred roller coasters had been built in the United States.

The Golden Age of Roller Coasters came to a sudden halt with the Great Depression. Within six years, eleven hundred amusement parks would be out of business. There was no time to recover before World

War II seized the nation's attention and many of the parks' young male customers as well.

After the war, the struggling parks, with their scary rides, found themselves out of sync with the new population. Their young customers had gone to war, returned, and settled down. These new families were now raising a crop of baby boomers, and according to park owners, they weren't coming to the parks because they either were transfixed by movies and television or were looking for outdoor fun that better suited their toddlers and preadolescents. In 1955, Walt Disney presented a new kind of park to appeal to them, one featuring quaint streets, jungle paths, enchanting castles, and frontier forts. Disneyland opened without a single ride based on falling, and certainly nothing as threatening as a roller coaster. Not having Disney's bankroll, the old-fashioned parks could hardly emulate Disney's extravagant formula for appealing to new tastes, and only a few limped along throughout the next decade.

By the early 1960s, the first of the baby boomers were reaching adolescence, and it seems they should have streamed back to amusement parks. They didn't. Instead, guided by the do-your-own-thing ethos of the era, teens would forsake packaged gravity thrills to seek out and control their own doses of falling, triggering an expanding proliferation of gravity recreation that continues to this day.

Surfing, which exploded in popularity after the release of the movie *Gidget*, set the mold for virtually every gravity sport since, establishing the board as the most common tool and rebellion as the requisite attitude. *(Copyright © Red Bull / Grant Ellis)*

## CHAPTER 7

# Young Gods Bronzed with Sunburn

We had our own style now. We didn't have to act like square football jocks. The creative freedom and exhilaration we'd found in surfing was affecting our whole lives. —Mike Doyle, recalling the birth of surfing culture in the 1960s

The rebirth of gravity play would begin with surfing. Surfing would have a galvanizing effect on a large part of American youth, beginning in the summer of 1962. On the radio, surf songs filled the airwaves: "Wipe Out" by the Surfaris, "Surf City" by Jan and Dean, "Surfin' USA" by the Beach Boys, and dozens more. I was growing up in southern California at the time, and although I was only ten years old in 1962, I instantly embraced surf culture, as did every kid I knew. We would hide *Surfer* magazine in our schoolbooks, stick Hobie surfboard decals on our binders, sketch surfers on huge waves on the back of our history notes. We wore Hang Ten shirts, which featured a wide "surfer stripe" across the chest. I ironed my wavy red hair in an attempt to make it fall over my forehead in the surfer's trademark slack locks. The fact that neither my friends nor I actually surfed didn't bother us; nor did we realize or care that we were absorbing only the commercial shadow of surfing. As far as we were concerned, surf culture was simply youth culture. Our culture.

Surfing would become the seminal gravity sport and would set the mold for what followed. Nearly every gravity sport since owes something to surfing, whether to its use of a board, its hedonistic culture, or its defiant attitude.

CONTRARY TO WHAT many teens at the time probably thought, surfing was not invented in southern California in the 1960s. Captain James Cook was the first Westerner to see surfing, shortly after he shocked Hawaiians by sailing over their horizon in 1778. In a detailed illustration of Cook's arrival in Kanakakooa Bay made by his artist, the ship's mast towers over the Hawaiians' outriggers; in the foreground is the first Western image of a man on a surfboard: he is on his stomach, paddling out to meet the strangers.

Later, it was the Hawaiians who shocked the explorers by demonstrating how they slid down huge waves on these planks of wood. "Their first object," wrote James King, a lieutenant on Cook's ship, "is to place themselves on the summit of the largest surge, by which they are driven along with amazing rapidity toward the shore. The boldness and address, with which we saw them perform these difficult and dangerous maneuvers, was altogether astonishing, and is scarce to be credited."[1]

The Hawaiians themselves couldn't tell Cook how old surfing was. Everyone had always surfed: men, women, and children; common folk and royalty. Surfing was woven into their oldest myths, such as the one about Kalamainu`u, who had a tongue that she removed and gave to her husband to use as a surfboard, or the story of Kolea-moku, who was betrothed to the ruling chief but instead eloped with her lover Kiha-a-Pillani after surfing with him for four days. There were competitions: in one, the victor won four thousand pigs and sixteen canoes. There were even prayers to bring good waves. "Arise! Arise, you great surfs from Kahiki, the powerful curling waves," went one. "Well up, long raging surf!"[2]

The Hawaiians told Cook about a legendary queen named Kaneamuna who loved to surf (and who also loved to slide: on a mountain near her village she had a slide built of smooth pebbles covered with rushes, and she spent hours sliding down on a wooden sled). Her surfboard, it was said, was buried in a cave after her death. The board was actually found in 1905, and when the story was researched, it was discovered that Kaneamuna had lived during the reign of King Keawenuiaumi, about the time of Shakespeare, in the early to mid-1600s. How far back surfing goes is still unknown; a good guess by scholars is that beginning about A.D. 400, while Goths were destroying Rome and Saxons were overrunning Britain, Pacific Islanders began to surf. Surfing, as Westerners came to learn, was widespread across a huge area and had been practiced for centuries in New Guinea, Easter Island, New Zealand, the Solomon Islands, the Western Carolines, the Vanuatu Islands, Fiji, Samoa, and Tahiti.

But the natives of Hawaii were on the losing end of eighteenth-century Western expansionism. In Cook's wake came traders, adventurers, and pirates, bringing diseases that killed Hawaiians in massive numbers: in 1778 Cook had sailed into a nation of a third of a million people; by 1893, there would be only forty thousand left. As the Western influence rose, Hawaiian royalty capitulated. In 1819, when King Kamehameha died, the government abandoned the Hawaiian set of religious codes, in which surfing was integral, and canceled the annual festival that celebrated native sports—including surfing.

The next year, the first missionaries arrived. They were Calvinists, among the least fun-loving Christians around, and once again, a gravity sport was attacked as the devil's work. In 1838, W. R. S. Ruschenberger wrote that the missionaries "state these sports to be expressly against the laws of God, and by a succession of reasoning, which may be readily traced, impress upon the minds of the chiefs and others the idea that all who practice them secure themselves the displeasure of offended heaven."[3] The Calvinists converted Kaahumanu, widow of the king and now acting ruler. Although she formerly had surfed herself, she now condemned the practice.

The suppression of surfing by missionaries was so blatant and cruel that Hiram Bingham, among the first clergymen to arrive, felt compelled to defend his church. "The decline and discontinuance of the use of the surfboard, as civilization advances," he wrote, "may be accounted for by the increase in modesty, industry and religion, without supposing, as some have affected to believe, that missionaries caused oppressive enactments against it."[4] Few believed him. By 1892 one anthropologist wrote, "Today it is hard to find a surfboard outside our museums and private collections."[5]

By 1900, the few surfers that remained had completely lost the old skills. They would lie down on their board most of the time; if they stood, it was to scoot straight down the face of the wave, like an outrigger sailing with the tide.

BUT THAT YEAR there appeared on the beach of Waikiki a seventeen-year-old boy named George Freeth. Freeth was half-Hawaiian and half-Irish; his mother was a member of the Hawaiian royal family; his father, a sea captain. It was the perfect heritage for what Freeth was about to do, which was to introduce the Hawaiian gift of surfing to the Western world.

Freeth first gained attention by reintroducing "wave-sliding," the art of cutting diagonally across the face of the wave to sustain the ride, and he reintroduced "stand-up" surfing as well. He also experimented with a more maneuverable board, cutting the traditional sixteen-foot board in half and then reshaping it. And because he looked white—almost Nordic—his presence encouraged whites to take up the sport, and an interracial surf club, the Waikiki Swimming Club, was formed in 1905.

When Jack London visited in 1907, he and his surf instructor, Alexander Hume Ford, walked from their Waikiki bungalow at the Hau Tree Inn down the sandy path to Kuhio Beach, where Freeth happened to be surfing. London was awestruck and published a long article that year in *Woman's Home Companion*, later adapted into a chapter of his widely read adventure book, *The Cruise of the Snark*, published in 1911. In it, he called surfing "a royal sport for the natural kings of earth," and described Freeth as "a man, erect, full-statured, not-struggling frantically in that wild movement, not buried and crushed and buffeted by those mighty monsters, but standing above them all, calm and superb, poised on the giddy summit, his feet buried in the churning foam, the salt smoke rising to his knees, and all the rest of him in the free air and flashing sunlight, and he is flying through the air, flying forward, flying fast as the surge on which he stands."[6]

At the end of that summer, Freeth, now twenty-four, moved to Redondo Beach, California, a land-boom development owned by Henry Huntington, who had stretched his railroad's lines there and who now wanted to sell land. He had also just opened the Pavilion, a three-story Moorish addition to his Redondo Hotel that covered 34,069 square feet, and "the world's largest saltwater plunge," to go with it. As a promotion to attract visitors, he hired Freeth to demonstrate surfing.

Freeth was the right man in the right place at the right time, as the southern California shoreline would become a mecca for gravity entertainment and recreation of all kinds. In 1904 at Venice Beach, Abbot Kinney had built a replica of the Italian city, complete with canals; he had also opened a "pleasure pier" that year. In 1911, Alexander Fraser would open his "Million Dollar Pier" next door. It would be followed by the Sunset Pier in 1921 and the Lick Pier in 1922, all of them eventually filled with gravity rides: fourteen roller coasters, including the legendary Big Dipper, along with toboggan rides and slides. On the Dragon Bamboo Slide at the Venice Amusement Pier, riders clung to a straw mat as they whipped down a ninety-eight-foot-high corkscrew. Riders sitting in the flat-bottomed boats atop Shoot the Shoots would plunge down a 120-foot runway to splash into a pool filled with 150,000 tons of water. It was also at Venice that the best show parachutists would perform. In 1911, Grant Morton would leap from a plane clutching his chute in his arms, and then float safely into the ocean—the first man to jump from a plane with a parachute. The West Coast's first official airport would be designated in 1914, established by movie mogul Thomas Ince in Venice as a base for his stunt pilots; and it was at Venice, decades later, that skateboarding would be transformed into an aerial sport.

But before these attractions, there was George Freeth. In 1907, Huntington billed him as "the man who walked on water." Abbot Kinney also hired Freeth; he bounced between Redondo and Venice, performing several times a day all summer and drawing thousands to see him. Adding to his acclaim was his work as one of the coast's first professional lifeguards. In 1908, when a violent squall hit Venice, Freeth almost single-handedly saved three boatfuls of fishermen threatened by the huge breakers. After rescuing men from two floundering boats, he dove off the end of the pier into the maelstrom, swam to a third capsized skiff, and held the men above water until a lifeboat arrived.

Freeth became an all-around watersports wonder, training Olympic swimmers, inventing the lifeguard's buoy still used today, and introducing water polo to the United States. He was the first to surf not only at Redondo and Venice but also at Balboa Beach near San Diego and Palos Verdes to the north. Everywhere he went, he found young men who wanted to learn to surf, and he taught them, seeding California for the surfing craze that would come years later.

Freeth, though, wouldn't live to see it. He died during the influenza epidemic of 1919.

THE MANTLE OF surfing ambassador next fell to Duke Kahanamoku, a friend of Freeth's from his Waikiki days. Named "Duke" after his father, a policeman, who was named for the Duke of Edinburgh following the Englishman's visit in 1869, Kahanamoku nevertheless was distantly related to Hawaiian royalty.

As a swimmer, Duke had already proved a prodigy. In 1911, when he was twenty-one, he entered his first competition in Hawaii, and although he had done no training, he broke the world's record for one hundred yards by four seconds, leaving his nearest rival thirty yards behind—and he had done it not in a calm swimming pool but in Honolulu Harbor. The Amateur Athletic Union didn't believe it. "What are you using for stopwatches? Alarm clocks?"[7] it replied to the news.

But it was true, as Duke proved by winning a gold medal in the event at the 1912 Stockholm Olympics (he would go on to win medals in 1920 and 1924). Europeans were smitten by this tall, exotic-looking, pure-bred Hawaiian who seemed to have a royal bearing that went with his name. He toured the continent; on his return to Hawaii he was greeted by the Royal Hawaiian Band and a cannon salute.

Duke was a swimming star, but audiences were fascinated by his surfing prowess. In 1912, he surfed for crowds in California at Balboa Beach and Corona del Mar, and he surfed at Atlantic City, New Jersey. "I paddled out beyond the last line of breakers," he recalled. "It was all of a half-mile [out] . . . I caught a giant swell and roared in all the way to shore at express train speed. The throng on shore loved it."[8] In 1914, he went to Australia. "I must have put on a show that more than trapped their fancy," he said later. "For the crowds on shore applauded me long and loud. . . . I soared and glided, drifted and side slipped, with that blending of flying and sailing which only experienced surfers can know and fully appreciate. The Aussies became instant converts."[9]

Converts were made wherever Duke surfed. At Waikiki, tourists at the developing resort crowded around him, and more of them were now riding the waves: the handful of local surfers at the beach in the early 1900s had grown to a few hundred by the 1920s. Freeth's surfing premiere in southern California had sparked a small band of surfers to form, and Duke's visits over the next forty years would steadily expand it. Progress was slow, but it was here, almost uniquely on the mainland, that progress was made. In the 1920s, "there were only six surfboards in the entire United States," recalled pioneering surfer Sam Reid, "and they were in Southern California."[10]

Although that was likely an exaggeration, it was true that no boards were commercially available; surfers made their own, and there weren't many surfers. This would begin to change when Tom Blake, a boy from Wisconsin who had been bumming around the country, landed in Santa Monica in 1921. He got a job as a swimming instructor and began to make a name as a new talent (he would set a world's record for the ten-mile open in 1922). That first year, he also stumbled across a surfboard at the swimming club where he worked—a ten-foot plank of redwood, the standard of the day—and took it out into the ocean. Within a few months, he had fallen in love with the sport.

After moving to Hawaii in 1926, he studied the century-old surfboards hanging on museum walls, particularly the olo style, a sixteen-foot model used only by royalty. By this time, Hawaiian surfers were using shorter boards—usually ten feet—mostly because even at that length the solid-wood boards could weigh a back-breaking hundred pounds. Blake discovered that the old boards were indeed solid wood but of a lighter variety that had fallen out of use: it was far easier for the new surfers to buy a plank of redwood from the lumberyard than to build boards from local trees. Convenience, more than suitability, guided surfboard design at the time.

Back in his workshop, Blake built a sixteen-foot olo board based on the ancient design, and then to lighten it, he drilled it full of holes, reducing what would have been a 150-pound board to a hundred pounds. He then glued a thin layer of wood veneer to the top and bottom to seal it, creating the world's first hollow surfboard. In 1928 he entered the first Pacific Coast Surfing Championships, held in Corona del Mar, which included a paddling surfboard race—an important event at the time. Carrying his enormous, cigar-shaped board, some in the crowd of ten thousand actually laughed at him. At the starting signal, he had trouble getting the long board up to speed and immediately fell thirty yards behind the pack. But once gliding, he caught up and pulled ahead. By the end of the 880-yard course, he had won—by one hundred yards. In a half-mile race two years later, he shaved more than two minutes off the old seven-minute record.

What was more remarkable was that by copying the old olo design, he discovered why the chiefs had made them so long (some were twenty-four feet). One traditional surfing spot for royalty had been Ka-lehua-wehe on

Oahu, but by Blake's time the spot had been all but abandoned. "It is a mile paddle from the Outrigger Club," Blake wrote later, "and with a short board the rider has to get dangerously near the break to catch these big waves." When he invited friends to surf there, Blake noted, "Excuses come thick and fast." The secret to catching the formidable surf was a lighter, longer board—like Blake's. With it, "the swells can be easily picked up, just as the ancients could do with their olo boards."[11] By 1930, Duke Kahanamoku, now a friend of Blake's, had lost much of his enthusiasm for surfing—that is, until he rode Ka-lehua-wehe with Blake on the new board. "The first big swell Duke caught went to his head like wine," Blake recalled. "He yelled and shouted at the top of his voice as he rode in."[12]

It wasn't that Blake's board was perfect for all surfing—it was hard to turn, for one thing—but by hollowing out the board and experimenting with length, Blake opened the eyes of board makers, and "imagination of design ran riot,"[13] as early California surfer Tom Reid remembered. No one was more innovative than Blake himself. In 1929, he patented a hollow board that used chambers instead of drilled holes, cutting the weight to sixty-five pounds; in 1931, he invented the sailboard, and that same year designed the first production surfboard, manufactured by the Thomas Rogers Company; in 1935, he added a fin to a surfboard for the first time.

Other designers took over from there. In 1938, John Kelly trimmed the bottom of a board's tail into a shallow V-shape, dubbed the "Hot Curl" board, which gave it better traction and allowed surfers to ride the huge, twenty-five-foot waves on Hawaii's North Shore for the first time. Eleven years later, Bob Simmons, a straight-A Caltech engineering student, Douglas Aircraft mathematician, and also an eccentric loner, showed what he could do with formal training and access to new military-developed materials. In 1949, he created the first styrofoam board covered with fiberglass and wood: it weighed only twenty-five pounds. He sold a hundred of them that summer. He developed a hydrofoil bottom, which made the board so fast he became famous for knocking over surfers who couldn't get out of his way. When demand for his boards overwhelmed him, he quit and fled to San Diego to live a hermit's existence on the beach, and died surfing there in 1954.

In the late fifties, Hobart "Hobie" Alter and Dave Sweet independently succeeded in molding surfboards from polyurethane foam. They were not only light—around thirty pounds—they also could be fiberglassed without the wood sheathing that fragile styrofoam required and could be easily mass-produced. In February 1960, Hobie sold 173 boards in two days.

By this time, surfers were acquiring a culture. It had sprung from the Waikiki Beach Boys, a loose fraternity of locals who were hired by the hotels to cater to guests during the beach's 1920s boom. They would teach surfing, pilot catamaran rides, fetch towels, and charm the guests—especially the women. After spending all day on the beach, they would gather at nightspots and got a reputation as carefree carousers and good dancers ("Beach Boys Sling 'Wicked Hoof' to Win Honolulu's First Charleston Contest,"[14] read one headline). They had nicknames: "Panama," "Boss," "Brains," "Indian," "Curly," "Steamboat," "Turkey," "Splash."

With Tom Blake, Duke, and others bouncing between Hawaii and California, the live-for-today surfing ethos was adopted by West Coast surfers, still a group elite enough so that one surfer meeting another on a beach would instantly strike up a conversation and usually a friendship—and may have known the other already by reputation. After all, in the 1930s, there were fewer than a hundred regular surfers in all of California.[15] After World War II, surfers, like the Beats, rejected the era's worker-bee conformity and were happy to eke out just enough of a living to surf. Californians visiting Hawaii would bunk down in rented shacks, sixteen to a room, and dine at all-you-can-eat restaurants, passing plates of food out the window to their hungry buddies.

WHILE ALL THIS was going on, the vast majority of the public still had little more idea of what surfing was than it did of kayaking. That was about to change, and change abruptly, beginning on a day in late June 1956, when a short stranger pulled a surfboard from a Buick convertible at Malibu beach. The stranger was spotted by three regular surfers, Mickey Muñoz, Mickey Dora, and Terry "Tubesteak" Tracy, who weren't about to have their beach invaded.

"Hey," Muñoz shouted, "Go back to the valley, you kook."[16]

Startled, the stranger dropped the board and stumbled. Tubesteak decided to go down the hill to help. Offering a hand, he blurted out, "For Chrissake, it's a midget. A girl midget. A goddamn gidget."

"I am *not* a gadget," the fifteen-year-old shouted. "My name is Kathryn and furthermore keep your filthy hands off me, you creep. And while you're at it, tell those two clowns up there to shove it."

The "gidget" was Kathy Kohner, whose father happened to be Frederick Kohner, an established screenwriter. He knew a good story when he heard one, and as Kathy fell in with the Malibu surfers, she came home with plenty of material. Frederick took the characters, setting, and some elements from her daughter's reports and concocted a light novel that was rapidly turned into a movie.

Released in 1959, *Gidget* is the story of a sixteen-year-old searching for romance among a band of surfers who instead treat her like a kid sister. The filmmakers were clearly trying to make a modest musical comedy, not the world's first surf movie, as *Gidget* has more in common with *Gigi* than with *Endless Summer*. One scene has surfer Moondoggie crooning Sinatra-style, backed by an off-screen string orchestra. Surfing is a backdrop; *Gidget* portrays surfing about as accurately as *South Pacific* portrays war. In several scenes a half-dozen surfers catch the same wave. Unnaturally crammed together to fit the camera's frame, they bump into each other and fall off their boards with some regularity.

What was probably most important to teens at the dawn of the 1960s was *Gidget*'s portrayal of the rebellious surfer lifestyle. Although the movie takes pains to refute it in the end, one of its main characters espouses a philosophy that foreshadows the hippies' call to nonaction. Kahuna, the leader of the surfers, is a self-declared surf bum who wants nothing to do with rules, work, or conventional society. Later he reforms and even takes a job. But the movie's young audience probably saw through this tacked-on ending, somehow divining through all the fictional layers of the movie and novel to connect with the attitude of the actual surfers who had inspired the story. Kahuna's ethos was, in fact, the prevailing philosophy of surfers, and *Gidget*'s audience seemed to know it. To them, Kahuna was probably still out there on a Malibu beach, defiantly jobless and surfing.

Although real surfers laughed at *Gidget*'s portrayal of their sport, the movie was an enormous hit. Its release is universally credited as the event that launched the surfing juggernaut into popular culture. Malibu surfers were soon complaining that hundreds of would-be surfers were overrunning their beach—and no wonder, as Hollywood continued to crank out beach movies that featured plenty of surfing, including *Gidget Goes Hawaiian* (1961), *Gidget Goes to Rome* (1963), *Beach Party* (1963), *Bikini Beach* (1964), *Ride the Wild Surf* (1964), *Beach Blanket Bingo* (1965), *How To Stuff A Wild Bikini* (1965), *The Beach Girls and the Monster* (1965), *Ghost in the Invisible Bikini* (1966), and *Don't Make Waves* (1967).

Pop music, especially in southern California, immediately caught the surfing wave. Among dozens of groups and hits there were the Beach Boys ("Surfin' USA"), Jan and Dean ("Surf City"), the Ventures ("Diamond Head"), the Chantays ("Pipeline"), the Surfaris ("Wipe Out"), the Bel Airs ("Mr. Moto"), the Pyramids ("Penetration"), and Dick Dale ("Miserlou"). At one point Dale's records occupied the top four slots among the top-ten records in Los Angeles. In 1961, twenty-one thousand fans stormed Dale's concert at the Los Angeles Sports arena, vying for only fifteen thousand seats. In Pasadena in 1962, Dale broke all attendance records, selling out the three-thousand-seat civic auditorium every weekend for a month, while a crowd of up to four thousand often danced in the streets outside. In 1964, when the Rolling Stones arrived for their first U.S. tour, they had to resign themselves to being the opening act for the headliner: the surf band the Trashmen, whose smash hit was "Surfin' Bird."

Teen fashion was also swept up in the surfing onslaught. Because commercial swim trunks of the time would rip out under the demands of the sport, a cottage industry had sprung up, with a few mom-and-pop operations making more durable trunks from canvas. After *Gidget*, the first surfwear company, Hang Ten, was formed; shortly after, swimwear giant Catalina put out a surfwear line, followed by Sears Roebuck and virtually every other clothing manufacturer who could get in on the trend.

Surfing as a craze didn't last long. By 1964, surf music had been eclipsed by the British invasion, and by 1967, the year of San Francisco's "Summer of Love," surf culture had been knocked from its youth culture pedestal by the hippie's influence—a lifestyle more suited to landlocked regions because it didn't require a beach.

But surfing's influence hardly evaporated. Surfing had announced that sports didn't have to be "square" and didn't have to be team oriented. It showed that gravity could easily deliver thrills bigger than hitting a home run or making a touchdown. Its conformity-be-damned philosophy seeded the hippie movement and informed virtually every youth subculture that would follow.

Moreover, surfing changed the way gravity challengers were viewed. Gone was the noble daredevil who challenged nature on behalf of his cheering audience. In his place was the archetype that has linked all gravity sports since: that of the defiant antihero.

## Dogtown

In 1967, Pacific Ocean Park, the last of the pleasure piers to hang on after the depression had wiped out the others, closed, defeated by Disneyland and by its own neighborhood in south Santa Monica, which had become a jungle of trash-strewn streets, weather-beaten shacks, and boarded-up buildings. The homeless loitered on corners, and in the gloom under the pier, men trolled for anonymous sex. People called it Dogtown, a name that was coined in derision but one that would become in a few years synonymous with a revolution in a new sport. It was in Dogtown that two skateboarders would add a huge wallop of falling to their sport by launching their boards into the air.

Tony Alva and Jay Adams knew Dogtown wasn't much, but it was still the beach, and for kids who didn't have a lot, it was the one scrap of the California dream they did have. As much as they could, they lived there. Jay's stepdad had a shop nearby where he rented anything a beachgoer might want, including surfboards, and at five years old Jay was already a pretty good surfer. Jay was a sidekick to Tony, who was twelve years old. They looked like brothers, their bushy blond hair falling to their shoulders, their bangs simply chopped off when they became too long to see through.

Together the boys had managed to work their way into the tribe of young men who surfed Dogtown by pelting nonlocal surfers with rocks to keep them away from the jealously guarded waves formed by the pier. For that, they were allowed to surf.

Immersed in surf culture, the two were keenly aware of the short-board revolution, as it would be called, taking place on the beaches around them. With all the changes in surfboard design, two things had remained relatively constant: the vast majority of boards were about ten feet long and three to four inches thick. A good ride was defined by cutting across the face of the wave and avoiding the curl; a great ride included "hanging ten"—putting your toes over the tip of the board. But in 1965, George Greenough built a kneeboard (a short board meant to be knelt on) with a thick tail that led to a flexible, spoon-shaped deck so thin it barely floated. To the bottom he affixed an eleven-inch flexible fin he designed. Riding it, Greenough could cut back and forth all over the wave.

Surfers who saw it realized that if the design worked for a kneeboard, something similar would work for a stand-up board, and in 1966, Nat Young showed up at the 1966 World Contest in San Diego with a board built by Bob McTavish that was nine feet, four inches long and only two and five-eighths inches thick and featured Greenough's radical fin. Young blew away the competition, playing with the curl and turning on the wave with a freedom that left the judges stunned.

Seeing Young win with his board inspired McTavish to push the envelope. The next year he paddled his latest board into the Australian surf: it was only seven feet, six inches long and weighed six pounds. The board performed as if on ball bearings. He could turn at a moment's notice, climb the wave to the curl, reverse back down the slope, then up again, then down, slicing across the slanted green water in tremendous arcs. Just down the coast from McTavish, Midget Farrelly had independently hit on the same idea.

And just down the coast from Jay and Tony's home, at Huntington Beach, David Nuuhiwa had also experimented with a short board, but not for a reason that had anything to do with improved performance. In 1968 Nuuhiwa was frustrated by a regulation that required surfers to quit the waters at 11:00 A.M., while kneeboarders—those with boards under four feet—could stay in as long as they liked. So Nuuhiwa cut a broken full-sized board to three feet, eleven inches, affixed a fin, and went out to take a stand-up ride in the midst of the kneeboarders. Surfers on the pier cheered as he rode in; the lifeguards stopped him, measured his board, and then reluctantly let him surf.[17] Although no more than a stunt, it made the point. Straight, stable rides and hanging ten were out. Carving up the face with swift cutbacks and steep, nearly airborne climbs to the crest were in.

Jay and Tony were swept up in the excitement, and they tried mightily to copy their heroes—but it was harder than it looked, and they failed more than they succeeded. When they became too frustrated, or when the surf was down, they would take to the streets on their homemade skateboards, crouching low, stretching out their arms, and pretending they were on the waves.

But skateboards had not enjoyed the design innovation that surfboards had. The skateboard had been around since the early 1900s, when kids took the wheels off roller skates and nailed them to two-by-fours to make scooters; if the pushbar fell off, it was a skateboard—although it wasn't called that. The first commercial skateboard was manufactured in 1959 and was nothing more than a toy: a hunk of wood with slippery metal wheels. In 1963, Larry Stevenson, publisher of *Surf Guide*, connected skateboarding to surfing by building his Makaha skateboards. Shaped like surfboards and fitted with clay wheels and adjustable trucks (the rubber-cushioned pivot that allows turning), Makaha skateboards were embraced by surfers and would-be surfers. Around the same time, Bill Richards was also building surfboard-inspired skateboards at his Val Surf Shop in North Hollywood. Although much improved from the metal-wheeled toys, these skateboards were still primitive: the clay wheels, meant for hardwood roller rinks, couldn't grip cement, and hitting a pebble the size of a Rice Krispie would stop the skateboard short and send the rider flying.

Nonetheless, Jay and Tony rode nearly every day. In 1970, the boys heard about a new place to skate: Paul Revere Junior High School in Brentwood. They rode their bikes up from Dogtown and immediately understood why the school had become a secret mecca for skaters. Built on a hill, its playground featured a fifteen-foot slope of smooth asphalt; in fact, there were similar slopes all over the school. Launching from the top, they could carve down these concrete waves, making the same moves a surfer would on a perfect breaker. And at Revere, the surf was always up.

Skating at Revere and on other banks they found, Tony and Jay pushed their skateboards to the limits, leaning into radical turns that put their bodies at steep angles to the cement, whipping around just in time to be on top of their boards, then spinning into a "360" before sliding back down the slope.

By 1973, when Jay was eleven and Tony eighteen, a design revolution had finally hit skateboarding. It had begun in 1970, when surfer Frank Nasworthy was nosing around the plastic manufacturing business of a friend's father. He noticed a barrel full of urethane roller-skate wheels that, it was explained, had been ordered by a chain of roller rinks. They had been made simply to be more durable for rental skates, but Nasworthy realized they would work on skateboards. Not only did they work, they also made the ride smooth and stable—and most importantly, they gripped the road like rubber. By the midseventies, Nasworthy's Cadillac Wheels or wheels like them were on virtually every skateboard made, including Jay's and Tony's. Their radical skating improved accordingly.

Tony and Jay, and all the Dogtowners, as they called themselves, knew they were good but had no idea how to compare themselves to the skaters in the outside world. What they saw in the magazines wasn't anything like what they did. At the time, the emerging professionals in skateboarding were divided into two camps: the freestylers and the slalom runners. The freestylers rolled across flat land, popping wheelies, turning 360s, hanging their toes off the front of the board, lying down on the board for a "coffin," or perhaps lifting themselves up into a handstand—most of the tricks adapted from old-style, long-board surfing. On the slalom course, skaters would whip down a slope around cones, like snow skiers. To the Dogtowners, it all seemed so lame. Where were the swift cutbacks and hard-leaning carves they saw short-board surfers pull off every day on the waves? Had these skateboarders actually *seen* anyone surfing lately? Yet if Tony and Jay wanted to be pros—and they did—they decided that they would at least have to join the game.

In 1975, they entered their first contest. They formed a team, the Z-boys, with the backing of the Zephyr Surf Shop, where they hung out and where they had found surfboard designers who took skateboards seriously and who created flexible skateboards that suited their style. Then they headed for the Bahane-Cadillac Skateboard Championship in Del Mar, the largest skateboarding competition held until that time, and the first to bring skateboarders from all over America together. And there, they blew minds.

Jay was the first of the Z-boys to skate in the freestyle event. From the instant he took to his board, the crowd knew something was different. Jay didn't push off for a gentle roll, the usual prelude before popping a wheelie or a 360. Jay shot out across the platform, pumping against the ground until he was speeding full tilt toward a drop. He crouched low, dipped, carved into the asphalt, slung his body this way and that. Just before he would have rolled off the drop, he spun his board in a turn so tight that he planted his hand for a pivot. Instantly, he had reversed direction, to shoot back into more quick turns. No one had seen anything like this. The crowd exploded. With each Z-boy that followed, the cheers rose, as dumbfounded competitors looked on.

In the following months, the skateboard magazines were filled with stories about the new style, and across America skaters began rolling down any slope they could find in the Dogtown style. Skateboarders would show up at the Zephyr shop to challenge Jay and Tony and the rest, to show them that they had learned what to do on a slope. But the Dogtowners were already ahead of them, again pushing skateboarding in a direction no one could have imagined.

Ironically, if the first inspiration of the Dogtowners had come from water, their next innovation came from the lack of it. In the mid-1970s, southern California experienced one of its most serious droughts, and the population was urged by officials to conserve water. Even the rich were doing their part; in backyards from Brentwood to Van Nuys, swimming pools stood empty.

And that intrigued the Dogtowners. They saw the virgin concrete in the pools as the perfect surface for their boards, the slope from the shallow end to the deep end custom-made for their style. Soon Jay was standing on the top of a friend's car as they drove slowly through the city's alleys, looking for empty pools whose owners weren't home. Word would get around about a new pool whose owner had a regular job, and soon it would be filled with trespassing Dogtowners, shooting across the shallow end, dipping down into the deep end, and then up the far wall as high as they dared. And shortly, they discovered an amazing thing: the only limit to how high you could go on the wall was how good you were.

It was risky business, pushing the limits of gravity and law at the same time, and it brought out the "local's only" ethic Tony and Jay knew well from their surfing days. "These kooks show up," Tony complained in an interview in *Skateboarder* magazine, "and try to follow our lines; they can't, and they eat it. . . . This one guy shows up with his friends at one of our spots and breaks his arm. It's no big thing; the guy could still walk, but he doesn't . . . he has his friends go call an ambulance. So we're yelling at this pussy, 'come on, we'll carry you out, be cool . . . you're OK; don't sweat it,' but no, he's laying on the ground crying, 'don't touch me; I don't want to be moved; it might be serious.' So the cops and the ambulance come and take the guy home to his mommy, and the next day the spot is totally busted . . . destroyed. Now this was serious. Just because this punk didn't take care of himself, we lose another good spot. Now we don't allow people who can't skate to ride our spots. We keep them out for their own good."[18]

In spite of the cops and the kooks, the Dogtowners again revolutionized their sport, and the two standouts were Jay and Tony. With a good push-off and a downward thrust on the hill, Tony could shoot up the wall to the pool's edge, spin on just the back wheels, and drop straight down to the pool's bottom. Still they pushed each other to do more. Jay flew up and let his front wheels fly over the edge before he turned.

Then, it happened: One day, Tony flew out of the pool altogether. He rolled fast down the pool's floor and headed up the wall, his body parallel to the ground. As he reached the lip, he crouched down and grabbed the board. He soared straight up into the air as all four wheels left the cement, now spinning free. In midair he turned, pointing his board downward. The wheels slammed back onto the wall; as he rode straight down, he flexed his legs to meet the curve of the pool's bottom, stood up, and rolled back to the shallow end.

"I was the first one to hit coping and then pull 'em up—I don't care what anyone else says," Tony stated in his second *Skateboarder* interview, and no one who has claimed otherwise since could make it stick. "You can ask Kent Senatore and Valdez and those guys, any of the guys that skated the Dog Bowl. I was the first one to use the coping as a launching trip to lift me up and out."[19]

Skateboarding had become airborne.

Yosemite Valley in the 1960s was home to a cult of wine-and-drug-fueled counter culture climbers who pushed far beyond what was thought possible. Typical was Warren Harding, shown here. Harding incensed other Valley regulars with his climb up the Wall of Early Morning Light on El Capitan because he drilled 330 bolts into the rock on the record-setting, month-long struggle to the top. *(Copyright © Galen Rowell / Mountain Light)*

CHAPTER 8

# Wall Rats

Because we're insane. —WARREN HARDING'S REPLY WHEN ASKED WHY HE AND DEAN CALDWELL CLIMBED YOSEMITE'S WALL OF EARLY MORNING LIGHT

IT WAS EARLY morning, November 3, 1970, but the Wall of Early Morning Light in Yosemite Valley wasn't lit. Normally the wall, which faces southeast into the gorge—the back, as it were, of El Capitan, the towering hunk of stone that stands like a three-thousand-foot gatepost at the valley's entrance—would have been glowing orange. But on that day, the weather was bad. Dark clouds pummeled the wall with freezing rain, as they had for days.

In the valley below, tourists at the Ahwahnee Hotel waited out the storm in leather chairs before the fire, looking out at drenched pine trees, sipping hot chocolate, flipping through *Life*. In a few weeks the magazine's pages would be filled with pictures of the two climbers who were just now shivering in hammocks slung on a few pins hammered into El Cap's sheer cliff. The coming *Life* article, along with an accompanying avalanche of publicity, would soon expose a cult of gravity junkies to millions of people for the first time, and the climb itself would reopen a wound that had been festering in the climbing community since the days of the extremists—a schism based on how willing a climber should be to fall.

At that moment, though, the two climbers on El Cap weren't thinking about the repercussions of their climb—they were wondering if they were going to finish it at all. They had been hanging against the flat granite, fifteen hundred feet up, legs cramping, for 107 hours. Endless hours that had withered body and soul, the hammock pinning their arms and legs to their sides, the downpour drenching their hopes. The rain flies weren't working. Water poured into their sleeping bags, the closest they had been to a shower in thirteen days. Everything, including them, was soaked, and everything stank. Their hands were swollen and raw from forcing their fingers into vertical cracks, twisting them into fleshy bulges to stick skin against rock, and pulling themselves up, a technique called jamming. Their knuckles were raw from fist-jamming big cracks, their palms scraped from hand-jamming smaller ones. In the thinnest, shallowest cracks they had hung their entire body weight on a few finger joints, and this had shredded their fingertips and cuticles.

Torn hands had been expected, but other things, such as the storm and the difficulty of the climb, hadn't been. After thirteen days, they were only halfway up. By leader Warren Harding's reckoning, they should have been on top by now instead of in this "wretched state of *soggification*," as he would later call it. "It soon became obvious that we had vastly underestimated the time that this venture would take us," Harding wrote. "Fortunately we had also greatly overestimated the amount of food and water required for a day's sustenance."[1]

If, at forty-six, the man who had been the first to climb El Cap's nose in 1958 thought he might be too old for this sort of thing, he didn't act it. Just as he had since the early 1950s, Harding had started this climb in defiance of good sense, even as much as that can be applied to the eccentric new breed of climbers who had made Yosemite their home.

Warren Harding lived at the foot of the cliffs farther down the valley, in Camp 4, a trash-strewn collection of cheap surplus army tents and stolen gear inhabited by clans of younger "wall rats." They had shaggy hair and a penchant for nicknames. Harding was "Batso"; his partner, Dean Caldwell, was "Wizard." Their clans, led by chieftains such as Harding, Jim "The Bird" Bridwell, and others, were "Bird's Boys," "The Vagrants," "The Stonemasters." They named their climbs *Lunatic Fringe, Crack of Despair, In Cold Blood, Meat Grinder.* Once again, the connection between gravity and counterculture leanings was confirmed: like the surfers who were living on gravity thrills and not much else, Yosemite climbers were resolute slackers. Most had no money beyond what they earned washing dishes or clearing trash in Yosemite Village, or sometimes, from recovering climbers' bodies, which would explode if they fell far enough. Sometimes it was hard to find all the pieces.

They bartered, fixing a car to get a mattress and a few homey items for their tents, which weren't so much for comfort as to lure women. They stole electricity to power the din of rock music by plugging into restroom light sockets. To eat, they often crashed religious picnics, putting up with the sermon until they could shove past the congregants and stuff themselves with fried chicken and pie. They drank jug wine and smoked marijuana when they could afford it, and they knew—as anybody who knew anything about climbing did—that they were the best rock climbers in the world.

In typical Camp 4 fashion, Harding had chosen this route while drunk. "Grandiosity of our plans," he noted later, "seemed to be directly proportional to the amount of booze we would consume at a sitting."[2] He had been relaxing in the meadow below El Capitan with a friend, drinking jug wine and peering up at the wall, trying to pick the ultimate Yosemite climb. He had seen a line that led to an utterly blank wall, and above that, at the top, the way blocked by a huge overhang of rock. It appeared, in a word, unclimbable. "The Big Motha Climb," he called it. A few weeks later, over a bottle of rum, he talked Caldwell into trying it.

Nothing was serious to these men except climbing, and that was very serious, even though they pretended it wasn't. "All summer we'd trained hard," Harding said later, "eating, drinking, loafing."[3] Actually, Harding had spent three months studying telephoto photographs that he had pieced together into a vertical map. They had arrived at the base with five haul bags—three hundred pounds of food and equipment. Much of the weight was in the hardware that would keep them from falling.

The system they used was standard for rock climbers. It works like this: at intervals, you wedge a piece of hardware into a crack—a piton, a nut on a cable, or a camlike device called a friend. If the face is blank, you drill a hole and screw in a bolt. To this you clip a gated metal ring called a carabiner, which looks something like an oversized, oval key ring. Once the carabiner is on, you slip the rope attached to your harness through it. Below you, your partner, who has secured himself against the cliff with hardware, wraps the rope around his body and stands ready to hold fast if you come off. This is called belaying.

As comforting as the system is in theory, in practice it's far more frightening. Although the equipment itself almost never fails, this happens occasionally. More likely is that you don't get the carabiner's gate locked shut and it comes off the hardware or releases your rope. Or much more commonly, you've botched the placement in the rock and the hardware rips out when you fall. Placing hardware is a mind-numbing, repetitive exercise, yet a climber's life quite literally hangs on every pin. The more you place, the safer you are, but stopping every few feet up a three-thousand-foot wall to hammer in a piton is unbelievably tedious, and a strain for your partner, who has to remove them as he follows. Hardware can be left in the wall (bolts always are), which saves time, but that means lugging far more equipment up the cliff. And a slow, safe pace has its own threat: you may run out of food far from the top.

An untried climb is a vertical wilderness. While the general route—up a natural crack, say—is envisioned, the particulars aren't known. You may plan to wedge a piece of hardware in a crack just above you, only to find you don't have one that fits that space or that your hold on the rock is so tenuous you can't possibly get a hand free to do the job. So you carry on a little farther. If you fall, you will fall more than twice the distance from your last anchor (due to rope stretch)—if you're ten feet from the last placement, you will fall more than twenty. Even if everything holds, the outcome isn't pleasant. On difficult stretches, the best climbers may risk roped falls of a hundred feet, which can slam them into the rock with a force that breaks bones. It is how most climbers are injured, and how some die.

Climbing an overhang offers a particularly sharp terror. Here, a fall will yank you straight down on any pins that have been pounded up into the overhang, popping them out and swinging you like a tetherball into the rock. If one holds, you're left dangling instead, staring down, perhaps, at a two-thousand-foot drop, with no way to get back on the wall. The overhang Harding and Caldwell would face was at the very top of El Cap, so the drop would be more like three thousand feet.

Because they had chosen a climb with so few holds, Harding and Caldwell had to place an enormous amount of hardware. By the climb's end, they had made the crucial decision of where to place a piece more than six hundred times—often making it when they were near exhaustion. Progress had been slow. There had been days when twelve hours of climbing had put them only ninety feet above where they had started.

By looking at Harding's photographs of the cliff, the pair knew they would have to cross a three-hundred-foot blank stretch of rock where there were no holds at all. And to get to the next climbable feature, they would have to cross it sideways. This meant that Harding would have to drill a hole, place a piece of hardware with a stirrup attached, hang in it, and then reach sideways as far as he could to drill the next hole. At best, this could only be a few feet. "Bolting vertically is tedious and strenuous enough, but bolting sideways is the utter shits," Harding said later. "Lean over to the left, tap-tap. Five or ten minutes later you have another rivet placed. Clip in and move about two feet."[4] Crossing the blank face took two days.

Each evening, the climbers had made camp, finding a spot to place hardware from which they could hang their enclosed hammocks, called bat tents. In these unsteady beds they could rest, eat, talk, or pose with their rears awkwardly outside the swinging bags to relieve themselves. But the storm that was now pounding down on them had trapped them in their last camp. Their bravado was fading. "I began to feel that maybe I wasn't going to make it." Harding confessed later. "A deep, dreadful feeling that death is not far away."[5]

Finally, the rain stopped. They managed to pull out dry clothes, shove their crumpled tents into the haul bags, and prepare to inch their way up again. The climbing was no easier. Five more days passed, another five hundred vertical feet. At that point they crawled up and onto a ledge to enjoy a moonlit dinner. It was only inches across but seemed a vast plain to them. The view was spectacular; the air, cold and clear. Two thousand feet below, the Merced River shone like a ribbon between the pines. The Cathedral Spires rose from the cliffs across from them; out farther, Taft Point towered. They unpacked their meal: French bread, cheese, salami, and a jug of Cabernet. Suddenly a voice through a bullhorn shouted up from the ground. "We've come to rescue you." Dumbfounded, Harding told them there was no need. Caldwell wrote a refusal, wrapped it around a stone, and threw it over. Then, above them, there was a commotion. A would-be rescuer was rappelling down the eight hundred feet from the top. Harding screamed at him to get away. They cursed and gave him the finger. Confused, he stopped, then pulled himself back up.

On the twenty-seventh day, they awoke knowing they were only sixty feet from the top. Harding affixed the rope to his harness and began to climb. He ran callused fingers over the wall, inches from his face, feeling for holds, concentrating. The overhang arched above; with each hold he clung closer to upside down. Here at last was the rounded lip. His belly scraped the granite, palms pulled, feet dangled then caught, pushed, and finally, up and over, he crawled up the slope of El Cap's top, three thousand feet of air at his back. It was his first look at flat ground in nearly a month—and it was covered with people. A mob of seventy reporters awaited him; thrusting microphones, cameras rolling. "I suddenly felt an overwhelming feeling of emotional release—sort of came 'unglued' for a moment,"[6] Harding wrote of the event. Caldwell arrived; together they gave interviews on the twenty-seven-day climb, the longest time anyone had stayed on a wall in Yosemite history.

Harding later claimed he was surprised by the reporters, but that remark may have been less than candid. He had long been the best source of his own legend, and he had friends in the valley who had primed the coverage a few weeks earlier. Not long after, the two climbers enthusiastically accepted invitations to appear on the *Merv Griffin Show*, the *Steve Allen Show*, and *Wide World of Sports.*

THE CONQUEST OF the Wall of Early Morning Light is famous—or infamous—in climbing circles, not because it shows the courage of the climbers but because many think it shows just the opposite. That was certainly the opinion among most of Yosemite's elite. Even before Harding had set foot back in Camp 4, clans were calling for his head.

It was an argument that dated back to the 1930s, when the extremists had dared to ascend the Eiger's face using ropes and pitons for climbing rather than just for safety. It had by now hardened into what can only be called a war of faiths. When climbers make judgements about what is good climbing and what is bad, they aren't usually talking about skill or the lack of it. They are talking about morality, and specifically they are talking about the degree of courage one shows in the face of falling.

As many saw it, Harding had sinned by so coveting a new route up El Cap that he created one that wasn't there. Instead of aiming for a natural crack or ridge, for example, he had deliberately chosen to confront a blank wall. The proof of his heresy was that he had used an outrageous 330 bolts, which covered 40 percent of the climb. Unlike other hardware, bolts can be used anywhere, not just in the natural creases of the rock, and they virtually never fail. In the ethics of orthodox climbing, to choose a route was to accept a covenant with rock and gravity and to accept the danger it brings. If there weren't enough holds, you were expected to either put your life at more risk or simply quit. Harding's answer was to bolt. In effect, Harding had cheated on the test given him by God: he had risked falling, but he hadn't risked it *enough.*

What the more traditional climbers admired most was sacrifice. They used the safety system, but they revered those who climbed in a way that denied its protection, climbers who would stretch out fifty or sixty feet before placing hardware and who would prefer a weaker pin to a solid bolt. "I have seen fine free and aid climbers pushing themselves to their limit using runners, nuts, pitons, and a rare bolt for protection," climber T. M. Herbert said, criticizing Harding and Caldwell's climb. "Many of my companions have risked nasty falls, even their lives, trying first ascents without placing a single bolt."[7]

At one time, even safe sleeping was frowned upon. "It became stylish," John Long recalls in his book *Rock Jocks, Wall Rats, and Hang Dogs*, "to see just how meager an anchor you could hang from for the night. Again, you were backed up by an absolutely bombproof, principal anchor, but what your hammock was actually slung from was as dicey as you were foolish."[8]

In the climbing community, Harding's apostasy ignited an international firestorm, similar to the extremist controversy but ironically with the establishment now firmly on the side of danger. "The question is," wrote the editor of *Mountain* magazine, Britain's leading climbing publication, "can we afford to stand aside and watch our sport being systematically denuded of its risk, character, and unpredictability?"[9] In Yosemite, there were shouting matches and fistfights; old friendships were destroyed. People called Harding a traitor, and he shot back that they were self-righteous "Valley Christians."[10]

If the Valley Christians had a high priest it was Royal Robbins, a lanky, intense man whose thick glasses made him look nerdish. His reputation as a climber, though, was supreme; it surpassed even Harding's. In 1951, Robbins had quit high school for the cliffs and the next year stunned veterans by free-climbing a treacherous route on southern California's Tahquitz Rock; that is, he climbed it without once using his rope or hardware to support his weight. At the time, it was the most difficult free climb ever done in the United States, and it had been done by a seventeen-year-old in high-top tennis shoes.

Unlike most of the Camp 4 crowd, Robbins was aloof and serious, almost an ascetic. He heard about Harding's climb, conferred with other leaders, and then spoke. "We thought it was an outrage, and that if a distinction between what is acceptable and what is not acceptable had to be made, then this was the time to make it."[11] The following January, he and Don Lauria started up Harding's route with a hammer and chisel. As they went, they chopped out every bolt that Harding and Caldwell had placed. But halfway up they changed their mind; they judged the difficulty of the climb, even with all the bolts, worthy, and they stopped chopping.

Harding later commented, "I don't give a rat's ass what Royal did with the route. . . . I can only assume that it was more of the evangelistic work that R. R. seems to feel called upon to indulge in."[12]

Yet Robbins made his point, and most climbers agreed: the more dangerous the climb, the better. It was inevitable that this opinion would inspire new climbers to push the risk of falling even further. Three years after Harding and Caldwell's climb, that risk was taken to an extraordinary new level.

## Ropeless

On a clear morning in 1973, Henry Barber stood peering up at Sentinel Rock, a three-thousand-foot tower of granite that stands across Yosemite Valley from Camp 4. Henry had seen it only once before, on his first trip here a year earlier. To the nineteen-year-old son of a prominent Boston banker, the Camp 4 crowd was unfathomable. Henry was not only conservative compared to the locals, he was out of place in the decade. Still living at home and enrolled in business college, he detested the aimless, drug-saturated lives of the typical Yosemite climber. "So many of them have slovenly behavior, no scruples, bad manners—little things that add up to an individual going nowhere," as he put it.[13] But he did respect their climbing. On his first visit he managed to impress them as well with his exceptionally clean style and by not violating the taboo of trying to make the first ascent of an unclimbed route, something the locals viewed as forbidden to outsiders.

Style was an obsession with Henry, an ethical point of pride that came naturally with his rigid personality. While the locals like Royal Robbins argued about how many bolts were proper on a climb, Henry didn't use bolts or pitons at all, because he believed that no matter how carefully they were pounded in, they damaged the rock for the next climber. Instead, he used nuts of varying sizes threaded with wire which he would slip into a crack until it jammed; to these he would attach his hardware, then carefully remove them when he moved on.

To the climbers in Yosemite, that put Henry on one side of a schism that had formed in the last few years between the older wall rats and the younger rock jocks. To the veterans, the only real climbs were multiday assaults on the big walls that included nights slung over a thousand-foot drop, and for these, pitons and bolts were obviously required. The rock jocks were a new breed who wanted to climb entirely free, using hardware for safety but not support. In his memoir, free-climber John Long recalled the taunts he used to get. "Fact is, Long," a wall rat had sneered, "you're a pussy. And a faggot. If you were half a man, you'd saddle up with us."[14]

Still, even the wall rats had to admit that Henry could climb, and he left that summer having made a good impression. Yet no one knew just how good he really was, which was the way he wanted it. Although he was becoming known back east for his climbing in the Shawangunks, Henry had kept his best climbs a secret, performed at obscure sites with one or two friends. It wasn't that he didn't want fame—he did, but on his own terms. On this, his second trip, he would likely find it: in a place as small as Yosemite, everyone knew what everyone else was doing, and they would know about his accomplishments soon enough.

Now he studied Sentinel. He knew its history well. Sentinel Rock had been a goal to climbers since the 1930s, when the easy routes were climbed. What remained was the north face, a sheer wall riddled with small cracks and perilous overhangs that rose eighteen hundred feet above a base of talus. The first team to try it in 1948 gave up after a hundred feet. That same year Jim Wilson and Phil Bettler reached only a little higher.

The following year Wilson and Bettler came back to launch a major attack with Allen Steck and Bill Long, two of the best climbers of the era, and plenty of rope, pitons, and a week's worth of food. It took the four the entire first day to claw their way up to an intimidating overhang, which they decided to leave until morning. The next day they nailed their way up so slowly that after fifty feet they realized they couldn't make it and retreated.

Undaunted, Steck and Long returned the next season. After two days of grueling work, they managed to reach the top of the Flying Buttress, an enormous pedestal that reached halfway up the cliff. And this is where they found real trouble: above them was a flared chimney—a slot in the rock that opened downward like the horn of a trumpet—which would be treacherous at best, but climbable. What didn't appear climbable was the hundred-and-fifty-foot headwall, utterly devoid of cracks, that loomed between the top of the buttress they were on and the beginning of the chimney. Again, they climbed down.

The next year, Steck was back again, this time with John Salathé. Salathé was, to put it simply, an eccentric. Five years earlier, at the age of forty-six, the Swiss blacksmith awoke in his California home one morning believing that an angel was speaking to him, telling him to change his life. He instantly became a vegetarian and shortly thereafter took up climbing. But this was climbing unlike anyone had seen before. Salathé, not particularly nimble, relied almost entirely on hardware to get him up a wall, yet he didn't seem to care much how safely the hardware was set. As a blacksmith, Salathé had improved on the old soft pitons with ones forged of a hard alloy, but he would place them in such weak spots that his climbing partners would be terrified to belay him, certain that the old man was going to die. It was nerve bordering on insanity—just what was needed on a climb like Sentinel's north wall.

Steck and Salathé took two days to reach the top of the buttress, where they camped for the night. The next morning, Salathé pounded a shaky piton into the headwall. He worked methodically and daringly, pounding pitons into little more than dimples in the rock and then standing on the slings he would attach to them. The wall was so blank he had to drill in six bolts. It took him ten hours to struggle up the headwall: an hour for every fifteen feet, and that only got them to the chimney, with nearly half of Sentinel still to climb.

What was worse, they were running out of water, and it was hot—a hundred and five degrees in the valley below. They spent another night. The next day they nailed their way up the chimney until it narrowed into a channel so tight that it was doubtful whether a climber's body would fit through it. Salathé wanted no part of it and climbed out over the sheer drop. Unfortunately, that way led to another overhang. Working his way along its underside, Salathé hung from pitons he had pounded into the ceiling that no one would have trusted but him. The sight scared the hell out of Steck. After that, the climbing got easier, yet it still took another day to complete: five days in all to conquer the north wall of Sentinel.

It was this wall, Henry Barber knew, that had helped make the reputation of his personal hero, Royal Robbins. In 1953 the eighteen-year-old Robbins, along with Don Wilson and Jerry Gallwas, pestered Steck to tell them the details of the route. Steck was vague, never dreaming that these nervy kids would try the wall's second ascent. But not only did they finish the climb, they cut three days off the five it had taken Steck and Salathé—an accomplishment that made the Valley crowd take note. Since then, the wall had been climbed many times along several routes. Once the main route became well known, it went faster, eventually becoming a one-day route. Royal Robbins, who had climbed it four times and knew it cold, eventually set the record for speed in 1961 with another world-class climber, Tom Frost: their time was an astounding three hours and fifteen minutes. By 1970, the route was climbed free by Steve Wunsch and Jim Erickson, who found a way around the notorious headwall.

Standing at the bottom of Sentinel, Henry Barber knew all this. He also knew that what he was about to do, if he succeeded, would make every one of these other climbs pale in comparison. Henry put his hands and a foot on the rock and lifted himself up; then he placed his other foot, then reached for his next handhold. He had studied the guidebook—even had it with him—but he had never climbed Sentinel before, so he didn't know exactly where he was going. Whatever way he got up, he was going alone, and he was going without hardware of any kind, and without a rope.

Henry had free-soloed before (as ropeless climbing is called); nearly every climber has started by clambering up a few rocks as a kid, and Henry had pushed this style further than most in his earlier years at the Shawangunks. He also knew the consequences. He once rushed to the side of an ambitious hiker who had tried to go up a cliff unroped and had fallen. Henry cradled the man's bashed and bloody head as he tried to dig the man's tongue out of his throat to clear an airspace. The hiker died in his arms.

And he knew the dangers of this particular climb. Jim Erickson, the man who had climbed it free, fell near the top but had been saved by his on-rope partner. Henry's friend Roger Parks hadn't been so lucky. Just a year earlier he had fallen on the route, also near the top, and was killed—even though he was roped.

Henry moved slowly over the first few yards, thrusting a hand into a crack here, planting his feet in a friction hold against the rock there. His body began to move into memorized positions culled from his repertoire; a shoulder stretched just right, a knee bent precisely. He saw the rock above like a mold he could flow into and also out of; up to the next move or if he had to, back down. Unlike roped climbers, a free soloist must know every move literally backward and forward.

Most soloists believe that safety lies in having climbed the route roped before. Henry believed just the opposite. "I first started soloing in the Shawangunks in the rain," Henry wrote later, explaining his philosophy. "I was worried about the wasps, and because it was wet and slippery, I had to be very cautious. . . . It made me start to realize that if you have no preconceptions about the route, you have little chance of getting heady."[15] The intense concentration helps guard against distractions—memories, thoughts of the future, the look of a lover's face. "The mind has to recognize when an inhibiting thought comes through and deal with it rationally. The mind has to be able to say 'look, body, we're not quite as attuned as we should be here; otherwise, that face would not have flashed.' "[16] Letting in a thought like that can kill.

Suddenly, climbing above the Flying Buttress, his flow was broken. The holds ran out. Here was a smooth face that would take desperate friction moves. Fear came in. He reached, stuck like a fly to the incline, but scared as hell. Another move, still scared, seven hundred feet of clear air below him. Another. The moves were working, but Henry was, as it is called, gripped. Finally, the slope lessened, the flow returned. He found a solid hold, checked his guidebook, and took five minutes to eat a candy bar.

Now he entered the chimney. The rock closing in around him felt comfortable at first, giving him a secure groove in which to wedge his body. But near the top it became claustrophobic; this is where Salathé had climbed out onto the face. Others, though, had climbed through the chimney, and that's the route Henry decided on. Squeezing his shoulders through the narrow gap, his arms practically pinned to his side, he wriggled up. Then something went wrong. He felt something slip; he felt himself falling.

Below him the chimney's throat threatened to drop him a hundred and fifty feet down the headwall to the buttress, then down eight hundred feet to the talus. He dropped suddenly through the widening slot, falling sideways. His elbow shot out against the rock. After sliding down three feet, he stopped, nearly upside down and held there by the friction of one arm. In this position he couldn't reach a hold with his other hand or with his feet. For fifteen minutes he sweated and wriggled against the granite, gingerly shifting his body fractions of an inch at a time. Finally, he scraped his way into an upright position and began to move carefully up through the hole.

After the narrows, the climbing was easier. The flow returned, and Henry moved through the last few problems of the climb with confidence. Finally, at the top, he sat down and looked below to the section that had killed Roger Parks. He felt like a voyeur; almost repulsed by the lewdness of it. His friend had died here. "I felt like a kid walking through the park and had some dirty old man come up to him and expose himself," Henry said later. "The little kid doesn't know what to do or say." Then he was utterly drained of emotion, completely blank. "There was no exhilaration in the completion of the climb, no thrill in having knocked off something no one had ever done before," he remembered. "It never occurred to me that the climb had meant a huge step forward."[17] Yet it had. Henry Barber, unroped and without knowing the way, had climbed one of the most frightening routes in Yosemite. When he looked at his watch he discovered something else: he had done it in two and a half hours, sheering three quarters of an hour off the speed record set by Robbins and Frost.

By eleven that morning, Henry was picking up his mail at the Valley post office when two climbers asked him if it was true. Word had evidently spread fast—Henry had told only a couple of people. When he got back to his tent at Camp 4, stunned climbers were already gathering around this nineteen-year-old kid from somewhere back east.

People had climbed big walls before; they had soloed before, they had climbed fast before, and of course, they had climbed unknown routes. No one, though, had ever done all these things at once. For all their big talk and macho strutting, neither the wall rats nor the rock jocks had ever conceived of a sustained confrontation with falling of this magnitude. Climbing, they suddenly knew, had abruptly and irrevocably changed.

Yet Henry Barber's accomplishment was more than just a milestone in climbing. It revealed the astonishing fact that humans, unaided by the use of tools—that which supposedly gives us our advantage over beasts—were among the best natural climbers in the world. Barber proved that we have the body for it, but more importantly, that with our unique ability to focus the power of the brain, a human could hold at bay the dread of gravity, the dread that, paradoxically, was the most likely force to send him hurtling to the ground. "Once you're scared," Henry has written, "the whole game starts to become very dangerous. In those situations there are quick subliminal flashes that are trying to grip you up, make you hold on tighter with your hands, things that make you all the more scared if you're scared already." But the mind's ability to focus on remembered moves can eliminate the fear. "As the rhythm develops, it carries me out of the doubts that might make me stop and say, 'What am I doing here? This is crazy.' " And the mind can also check the overconfidence that might then result. "At this point it's critical to have some distance on the situation, so that you can stop and say, 'I should not be doing this. I've got to get a grip on myself and work out this situation.' "[18]

With learned techniques and the capacity for sustained mental discipline, a human could climb a wall with the tiniest of holds, and climb it far higher than was ever imagined. By conquering the fear of falling, a human could conquer falling itself.

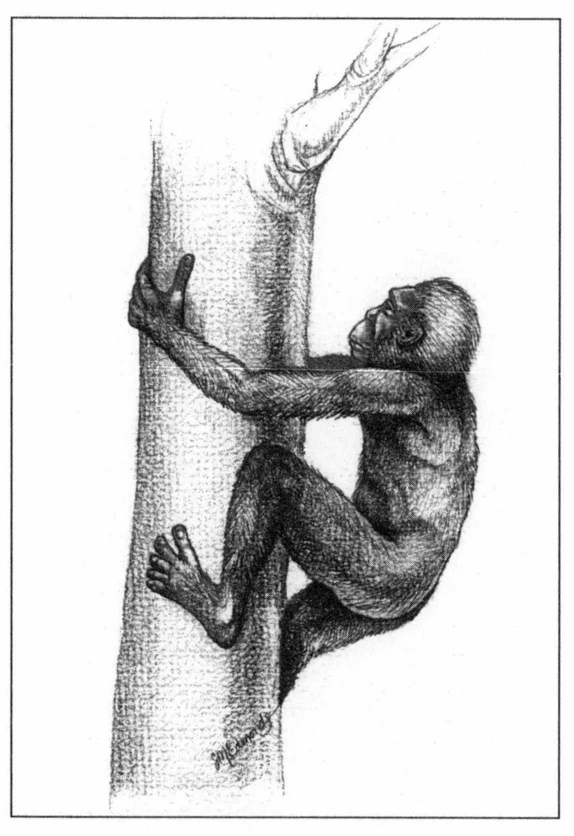

Although narratives of human evolution tend to emphasize the six-million-year era since our ancestors left the trees, the previous sixty-five million years arguably had a greater impact on human nature. During those eons in the rainforest canopy, everything from the shape of our body and limbs to color vision was formed as a direct response to the threat of gravity. Even our first known bipedal ancestor, *Australopithecus* (shown here; popularly known as Lucy), probably climbed often, ensuring that our unique relationship to gravity would remain intact. *(Copyright © Suzanne Edmonds)*

CHAPTER 9

# The Arborealists

> While broad aspects of our bodies place us in the mammals, so detailed anatomy collectively identifies us with those agile mammals, primates—the order that includes the lemurs, monkeys, and apes. Most primate features seemingly evolved for life spent mostly up in trees.
>
> —DAVID LAMBERT, *THE FIELD GUIDE TO EARLY MAN*

THAT OUR FEAR of falling is inborn is obvious: people don't have to be taught to stay away from dangerous drops. Yet science didn't get around to confirming this fact until the late 1950s, when a psychologist thought of a way to discover how early the fear develops in humans. Not coincidentally, she was inspired while enjoying a picnic on the rim of the Grand Canyon.

Psychologist Eleanor Gibson was taking in the canyon's panorama when she began to wonder if a baby, left unattended, would crawl over the edge. Back in the lab, she and fellow researcher Richard Walk devised a clever experiment to find out. What they created came to be known as a visual cliff: a checkerboard floor that led to a dropoff covered with strong glass. Would a baby venture out over the drop?

Gibson and Walk tested thirty-six babies between six and fourteen months old and found that all but eight refused to cross the glass, even when urged on by a parent. Some actually backed away from the cliff; others burst into tears at the gap between them and their mothers.[1]

While it's not surprising to find that Gibson and Walk's research confirmed that our fear of falling is inborn, we also know that the pleasure of falling starts when we're very young. Kids climb and jump almost as soon they can walk, and even before that, babies enjoy being tossed in the air. This fact is so apparent that it too has mostly escaped scientific scrutiny, although Robert Grossman, a researcher in perception, did write in *Perception in Everyday Life*, about how he used the sensation of falling to quiet a baby:

> I decided to give him a short bit of free-fall without letting my hands leave the surface of his body. He stopped all of his movement and looked vaguely at a spot far in the distance as if he were "checking out" this new experience. The baby was too young to laugh or show overt signs of enjoyment, but he didn't show any signs of being bothered so I tried again. He appeared to find the experience interesting, if not exciting.
>
> If the child does perceive this type of stimulation as exciting, then small amounts of gentle free-fall should be more effective in stopping a child from crying than rocking. Indeed, I have found this to be the case. If the child is somewhat older, he usually enjoys it so much that he begins to laugh."[2]

The paradox is that we're born with two conflicting instincts: one designed to protect us from gravity, the other urging us to play with it. What's most puzzling is why we would need the second instinct at all. As land animals, it seems we would be better served if we just stayed away from falling altogether.

The reason we don't is due to a simple fact: we aren't truly land animals. We're about as much a land animal as a duck is a sea creature. Like the duck, which adapted to water only after spending millions of years evolving flight, the human line took to the ground only after aeons evolving an extraordinary ability for an entirely different environment. That ability is acrobatics, and its birthplace was the trees.

This fact isn't emphasized by anthropologists, because to them it's a given and because they're more interested in the time when our ancestors began walking upright. After all, it was then that we acquired the feature we're most proud of: a big brain. But evolution is a long, unbroken chain, and while our recent ancestors were on the ground for about five million years, *their* ancestors were in the trees more than ten times longer. The truth is we are only recent immigrants to this new flat world.

Most of what we are—our body shape, our senses, even the basic circuitry of our brain—was formed over that vast stretch of time that our forebears spent among the branches. And it is there, in the rainforest canopy, that the answer to the question of why we both fear and desire falling can be found. In that world, falling wasn't just a danger. It was also—in the form of *controlled* falling—a necessity.

## Gifts from the Trees

The creatures who began our line, the line of primates, were latecomers to the trees; the earliest animals and plants had been aiming skyward since long before. The first plants crept from the sea onto land about 425 million years ago, gradually spreading outward to gather light. As the shore became crowded, plants began to grow taller to catch more of the sun's rays, literally overshadowing their competition. By 350 million years ago, giant ferns and primitive trees shot up a hundred feet or more and covered large parts of the earth with vast, dense forests. As the trees went up, so did the rest of life; among the first climbers were spiders, scorpions, and six-inch cockroaches.

Then a hundred million years later, things changed dramatically: the earth began to cool, and the forests shrank, forcing creatures to crowd into the warm pockets that were left and pressuring more of them to scramble into the branches for food. Reptiles took to the trees, and shortly after, plants began to flower and bear fruit for the first time. Now, packed with life, the rainforest canopy became a hothouse of new species. By the late Cretaceous era, about eighty million years ago, it was the richest depository of food and life the world had ever known, an environment more fertile than the land itself.

In spite of the treetop abundance, mammals, which had already existed for millions of years, remained scampering ground dwellers, kept in their place by the reptiles and dinosaurs who long before had filled nearly all the niches in the land, sea, and trees. Then without warning, the world suddenly changed again, this time the result, appropriately enough, of something that fell: an asteroid about the size of Manhattan.

It flamed through the sky sixty-five million years ago, hit the ocean near the Caribbean, and exploded into an immense cloud that choked the air for decades. That triggered a change (which still isn't completely understood) that began killing scores of species. As they died, new worlds were left wide open to mammals. Mammals went everywhere, and each place they invaded transformed them into wholly new animals.

Exactly why the founder of the primate line first climbed into the trees is unknown, as is its identity. A likely candidate is a small ground-dwelling creature known as *Purgatorius.* It looked almost exactly like a rat; virtually all mammals at the time did, and *Purgatorius* probably wasn't an especially successful model. It might have climbed because it was failing to compete against the more efficient ground mammals that would later evolve into true rodents. *Purgatorius* differed little from them except for two things. The first was that its brain was slightly larger for its body weight, a minor distinction at that point. The second seems even less significant—and yet it would completely remake the animal and in the course of aeons, give rise to a new animal that would remake the world.

## The Power of the Grip

That difference was in its hand. Its hand wasn't advanced; it was primitive—any hand was. Something like a hand formed on the first land vertebrates, the labyrinthodonts, salamander-like creatures who evolved from fish and crawled from the ocean 250 million years earlier. In nearly all land animals, the hand would disappear completely, giving way to something better suited for covering ground or tearing into prey. It had already morphed into a flat foot for *Stegosaurus* and a claw for *Tyrannosaurus rex.* With the beginning of the mammalian explosion, the power of the land soon molded land mammals in the same way. One of the ratlike mammals that migrated to the newly vacated plains, for example, began to push off more with one toe of its paw than with the others. As the animal grew larger, its favored digit became stronger, eventually growing as fat as its wrist while its other toes vanished. Today its descendent, the horse, gallops on four single toes.

For the ground mammals that would become rodents, the hand had already become more pawlike; for *Purgatorius*, it had not. Its hand was a little broader, perhaps; with one digit that diverged just a bit more from the others. When it first took to the trees, this didn't make much difference, because like other mammals who invaded the canopy, it climbed by digging its claws into the bark. But as the trees filled with competition, its slightly wider hand caused *Purgatorius* to do something other tree dwellers didn't: it began to grip.

This climbing technique gave *Purgatorius* a slight edge, and soon gripping began to transform *Purgatorius*'s offspring into completely different animals: these were the first primates. Over twenty million years, they gripped more and clawed less. Gripping forced their palms to slowly widen, their thumbs to move away from their lengthening fingers, and their claws to flatten into fingernails. Reaching for a grip stretched their arms and legs and loosened their joints. Primate forearms began to twist to allow the hand to face up or down. Its neck changed so it could swivel, to survey all climbing options (today's tarsiers, for example, can spin their heads around and look directly backward without bothering to turn their body). And its shoulders were entirely rebuilt: in land animals, shoulders only flexed enough to let legs move through a stride, but primates pushed their shoulders in every direction until the joint became a ball in a loose socket. Eventually, primates could swing their arms in a full circle—unheard of in a land animal. To keep track of its more mobile limbs, the primates' proprioceptive sense—the feedback system of nerves that monitor joint position—became finely calibrated to signal its brain of each limb's precise orientation in space. The rat body was now gone, and in its place was a new animal with a body flexible enough to sit, stand, bend, twist sideways, curl up, and leap through the precarious tangle of the rainforest canopy.

Not all primates developed all the abilities described here; many stopped at a particular specialization that served them. Swinging, for example, which is called brachiation by primatologists, is a form of movement that separates monkeys and apes from the line that led to tarsiers, sloths, and dozens of others. But nearly every improvement that made primates more acrobatic lies along our own line of origin, some surprisingly so: our arms are longer relative to our trunk than are a monkey's arms; our forearms can twist 180 degrees, while a monkey's can only turn 90 degrees; and our wrists are more mobile. Under the demands of the grasping hand, the primate body was created, and it is this body that we have largely inherited.

## Balance

Agility, though, came at a price, and that price was an increased risk of falling. To guard against it, the primate's vestibular organ, which senses balance, had to be refined.

Like the hand, the vestibular organ in vertebrates had evolved early. It consists of three connected semicircular tubes lodged inside the animal's head, arranged to sense movement in each of the three dimensions: one tube is horizontal, one stands upright facing forward, and the third stands upright facing backward. Liquid inside the tubes ebbs and flows according to the pull of gravity, and this movement is sensed by nerve endings and communicated to the brain. It's a system that works well in any environment (even fish have it), but like any other part of an animal's body, it has evolved in different species according to need. Today's lamprey, for example, is a sluggish fish that spends virtually all its time on a river or pond bottom; since it has little reason for a sharp sense of up and down, one of its vestibular tubes has disappeared completely.

For primates, the problem was exactly the opposite: rapid three-dimensional movement became their specialty, and as a result, the vestibular system that humans have inherited is unusually sensitive and complex.

Inside our vestibular tubes are two different systems that sense two kinds of movement. First there are the maculae, two gelatinous blobs, each about the size of a sesame seed, that are stuck to the inside of the tubes. One rests flat on the bottom of a tube, while the other is affixed to the tube's vertical wall. Embedded in each macula are hairs that bend when it shifts. If you tilt your head from side to side, the macula on the tube's bottom leans, the hairs bend, and nerves at their base signal your brain. They also react if you move forward or backward suddenly, as in a speeding car, for example.

The macula stuck to the tube's vertical wall records up-and-down movement. If you stand up suddenly, it is forced down; if you drop, it becomes weightless; that's one way you know you're falling. Like the bubbles in a carpenter's level, these tiny, constantly shifting bits of gelatin tell you directly about the pull of gravity.

But there is more to tell, including in which direction and how fast you're moving. This is handled by the vestibular organ's second system. Projecting into each tube's liquid are two more hair-studded blobs called the cupulae. Because the liquid and the cupulae have the same specific gravity, inertia keeps them from moving if you tilt your head or jump up and down. What gets the liquid to move is sustained turning. When you spin, for example, the liquid tends to lag behind, similar to what happens if you take a glass with ice and water and twist it quickly: the glass turns, the ice and water lag behind. If your head continues to spin though, the liquid in the tubes begins to spin too, washing over the cupulae, bending the hairs, and signaling to your brain how fast you're spinning and in what direction. Since the tubes run along three axes roughly perpendicular to each other, your brain can compare the signals and get a good sense of the precise orientation of your spin in three dimensions.

## New Eyes

With a gripping hand, flexible body, and a good sense of balance, primates were poised to travel through the branches more efficiently than any mammal had before. Yet to truly master their world, they needed something more: the ability to see where they were going more precisely—and more quickly. And so, in one of the most dramatic changes to their anatomy, the threat of falling transformed primate eyes.

The first primates had eyes on the side of their head, indicative of their status as prey that were best served by being alert to ambushing predators. The carnivores who ate them had eyes facing forward, for the chase; up to this point in evolution, that was the primary reason for binocular vision: to hold in focus the prey they were running down. But as falling became a greater danger to primates than attack, their eyes began to migrate from the side of the head to the face—and for the first time, they began to see depth. This was crucial: misjudging the distance to a branch by only an inch could prove fatal.

To wring the most out of its stereoscopic vision, the primate brain learned to judge distance not just by the disparity between the two images—this method is really only effective at close range—but also by sensing how far eye muscles pulled together or apart to focus on an object. Primates' depth perception became keener even than most predators (more so than cats, for example). And eventually, primate eyes began to see something no other mammals ever see: color. Now they could more clearly see red fruit high above, standing out against the leaves, and the sturdy brown branch hiding in a mass of green vines that would take them there.

Primate eyes—our eyes—became extraordinarily quick at seeing. They became coordinated to shift left or right, up or down, in exact tandem, pulling together or apart instantly to see near or far, always focusing the image on the most sensitive part of the retina, the exact center. Humans can assess visual input in an astoundingly short time; between 150 and 300 milliseconds. This is, in fact, the only way we see, taking in snapshots of the world with short flicks of the eye. Even when we stare at something, there is a microtremor of the eyes that keeps them moving.

To keep the image steady as primates leapt and swung, a direct link formed between the vestibular organ and the eyes, resulting in the vestibulo-ocular reflex, which is why you can keep your eyes focused on a stationary object as you turn your head. (This reflex is also why the world seems to continue to spin after you've whirled around too much: your eyes continue to move because the spinning liquid in your vestibular system is telling them to.)

With more signals pouring in from the primate's expanded senses, processing them now became an extraordinarily complex job for the brain—especially within the time a falling or swinging animal needed to react, which was practically no time at all. To work efficiently, most reactions would need to be carried out automatically by reflexes, which in any vertebrate are handled by the cerebellum, the lower brain.

The cerebellum gathers reports from the senses and sends out orders to a motor cortex, which in turn sends out packets of programmed movement faster than consciousness can understand them; this is why animals don't have to think about walking in order to walk. Primates, though, needed packets that went far beyond walking or running; they needed a huge repertoire of movements that could deal with any fix the animal might find itself in while climbing, swinging, or leaping through the trees. And as the number of programmed packets increased, the primate brain grew larger, in almost direct proportion to its burgeoning agility.

## The Birth of Consciousness?

It's no wonder, then, that we like the feeling of controlled falling: nearly every change that natural selection imposed over sixty-five million years of our evolution was meant to enable us to sail through the trees as easily as a fish swims through water. From our earliest origins, controlled falling must have been a deeply satisfying experience.

But there is another point to consider, one that gets to the core of how our dance with gravity continued to influence not just how we feel during a fall but what we think about it. That point is how much movement through the trees was to be controlled by raw instinct and reflexes and how much by deliberation. On this question, monkeys went one way and the great apes went another. What developed in the apes and in us, if primatologist Daniel Povinelli and physical anthropologist John Cant are right, is consciousness.[3]

According to their clambering theory, long after our line of primates developed the reflexes to maneuver through the trees, those reflexes began to fail at least part of the time. They failed because as our ancestors got larger, the automatic responses that the cerebellum produced, while good enough to keep a monkey from falling to the ground, weren't good enough to protect something approaching the size of a human.

Povinelli came up with his theory by noticing the difference between the way monkeys and orangutans navigate the canopy. Because monkeys are small, most branches in a tree will support them. They travel by jumping, swinging, leaping, and grabbing whatever's handy. Chances are, it will hold. If it doesn't, something nearby on the way down will. It's possible for a monkey to get away with this strategy because there is relatively more surface area to weight in a small animal than in a large one, and this means that air resistance is a much greater factor in slowing a small animal's fall. This is why a cat can fall from a height five times its body length without incident, while a human can't. As a consequence, a monkey has a much better chance of escaping injury even if it does fall to the ground.

For a great ape, like an orangutan, which can weigh up to 180 pounds, the trees are not so sturdy, and a fall is serious business. So orangutans clamber: they walk carefully, test their holds, even judge how to bend a branch enough so that it will gently lower them to where they want to go, like an arboreal elevator. Povinelli hypothesizes that this careful movement is what the common ancestor of the great apes and humans learned, and that by doing this, it began to understand that it was an individual who could cause things to happen, that it could make plans and reap benefits. In short, it became self-conscious.

Tests of cognition bear this out. When animals marked with a red dye on their forehead are put before a mirror, most act as if their reflection has nothing to do with them. A monkey will screech at the mirror to scare off the invader. Only our closest relatives—an orangutan, chimpanzee, or gorilla—will recognize itself and will prove it by touching the red spot on its forehead and then looking at its finger to see, perhaps, if it's bleeding. Even in the late stages of our evolution as primates, then, the threat of falling was pushing the qualities we call human into being.

## To the Ground

Long after primates had achieved mastery in the trees, their home began to change. In eastern Africa, between twenty and ten million years ago, the earth lifted nine thousand feet in two great bulges, forming the Rift Valley and cutting off rain to nearly half the continent. Apes on the dry side saw the forest canopy, their world for millions of years, shrink into pockets. There was little choice but to begin to adapt to a savanna dotted with only islands of trees—and that brought about the extraordinary change of walking on two legs.

To understand why it was extraordinary, you need to know what primates typically do when they travel across land. Today's baboons live primarily on the ground, and chimpanzees and gorillas spend much of their time there. Like the great majority of land mammals, these primates use all four of their limbs as legs. Baboons walk on the flats of their hands, and their arms and legs have evolved to be almost equal in length; in fact, their build and gait is nearly doglike. On the ground, chimpanzees and gorillas also use their arms as legs, walking on their knuckles. But for reasons that are still subject to a near violent debate among anthropologists, our ancestors didn't take this obvious path. Instead, they began to stand up.

From the beginning, though—and this is where much of the mystery lies—they must not have stood as inefficiently as an ape, or they wouldn't have pursued the stance. An ape stands unsteadily, ready to drop to all fours whenever it needs to cover ground quickly, because its build makes walking on two legs absurdly awkward. Its feet, which are more like hands, don't give it a stable platform. The sockets of its pelvis turn outward, forcing the ape to walk with its knees bent in a bow-legged waddle. Its shoulders are narrow and its chest is wide, so its arms, which are long to begin with, can't swing forward and back as they must for a steady gait. Dealing with all this, a chimpanzee trying to walk upright will expend more than twice the energy that a human does. Although our ancestors probably had many of these same problems, they must have already developed a more upright method of traveling through the trees, probably limb-walking, that made standing on the ground a better option from the beginning.

The irony is that after millions of years of evolution that emphasized balance and acrobatics, all apes who moved to the land left those skills in the trees and knuckle-walked or waddled to get around—all apes except one. Our ancestors did not. And because they stood up, their bodies began to adapt to a stance from which there was no way to relax. Walking permanently on two feet, they developed a constant, articulated effort against gravity.

Gravity remade this primate's body once again, but this time keeping nearly everything useful that had come from the trees and adding just what was needed for a new kind of locomotion. These animals were becoming us: our foot became a stable platform, with a big toe to push off against. Our pelvis changed to align our legs for striding forward; our shoulders broadened so we wouldn't have the ape's problem with swinging arms. Our spine straightened and became stronger (although it's still not strong enough, which is why so many of us have back problems). And finally, our sense of balance (the vestibular organ) was given a final boost.

By studying the skulls of earliest upright walkers, estimating the size and shape of their vestibular systems, and comparing them to both modern apes and humans, paleontologists Fred Spoor, Bernard Wood, and Frans Zonneveld have been able to trace this improvement.[4] They found that as we evolved on the ground, our horizontal tube shrank slightly, while the vertical tubes became enlarged. We needed less sensitivity to side-to-side movement but a lot more for front-to-back movement. With this equipment, we would not just stand—after a few more aeons, we would run, balancing on one foot at a time at speed, a trick of equilibrium and coordination unequaled by any other primate.

This is where the story of human evolution traditionally begins, with bipedalism, because this is where the line leading directly to humans becomes identifiably distinct. Yet it's only distinct because there's a gap in the fossil record. We know there must have been a gradual transition from tree-dwelling apes to two-footed walkers, but we haven't yet found fossils that show this evolution. What we have instead is a nearly complete skeleton of a creature that was already upright, *Australopithecus afarensis,* discovered by Donald Johanson in 1974, which he nicknamed Lucy. Lucy was clearly a walker; her hips, legs, and feet all bear ample testimony. And four years after Johanson's find, footprints attributed to her species were uncovered on a fossilized beach. Together, this evidence conjured up a vision of our earliest ancestor striding bravely out across the land, single file, traveling miles to find water or a patch of berries.

Yet even at this late stage of bipedalism, the image we have of Lucy and her kin as a fully committed land animal, free of the influence of falling, is almost certainly wrong. Lucy did walk on the savanna, but a closer look at her world shows that she could not have walked very far from shelter. The savanna then, as it is now, was filled with predators, and there's plenty of evidence that Lucy's kind was among the prey. In southern Africa, virtually all the remains of a later australopithecine—this one nearly as large as today's humans and presumably less vulnerable than Lucy—came from caves where these primates had been dragged by big cats to be eaten. One skull even retains the teeth marks of a leopard that fit with a cat's gruesome habit of hauling a carcass by the head.

Many anthropologists now believe that as our weaponless ancestors ventured out across Africa's grasslands, they could have had only one defense: the one they had already been exquisitely evolved to exploit—and that was to climb a tree faster and higher than what was chasing them.

In the excitement of discovering the earliest bipedal hominid, the features of Lucy's anatomy that told this story were at first overlooked. Once reexamined, it became clear that Lucy was still well built for climbing. Think of a chimpanzee that stood on two feet, and you have a good idea of her size and appearance. Her legs were short, her feet big, and her toes long and nearly finger-like. Her ankles were more flexible than ours, letting her bend her feet upward at a much steeper angle. She had upward-tilting shoulder sockets (ours tilt downward), long arms, long curved fingers, and powerful wrists. A good guess is that when Lucy was in trouble, she shot up a tree in an accelerated version of the way a Pacific Islander climbs a palm tree, by wrapping her hands around the back of the trunk and shinnying up with her feet on the sides. In a bit of prehistoric vengeance, some theories hold that her species also climbed regularly to scavenge the kills of the cats who would store their prey among low branches. At night, Lucy and her kind almost certainly slept in trees, as nearly all today's primates who have adapted to the ground still do (except gorillas, who, as the joke goes, sleep wherever they like). In other words, for perhaps another three million years after bipedalism began, trees were still home, and falling out of one was still a disaster: if the fall didn't kill them, then waiting jaws would. There was good reason for our acrobatic abilities to persist, and persist they have.

All in all, falling—avoiding it and controlling it—has had a profound effect on human evolution. If we hadn't taken to the trees in the first place, we might have evolved into the rat's cousin. If we hadn't stood up against gravity on two feet, we might now be the baboon's sibling. But under the influence of falling—both in the trees and later on land—we became primates, different from all other mammals, and then Hominidae, different from all other primates. Our grasping hand led to binocular eyes and quick brains, planting the seeds for tool making, reading, and abstract reasoning. Looked at this way, many of our cherished intellectual abilities are just a fantastic side effect of trying not to fall out of a tree.

And while we usually think of other species as having all the physical talent—they run faster, swim better, leap higher—we forget this remarkable fact: our unique evolution has made us the most versatile acrobats the world has ever known. Apes are certainly more nimble in trees, but if all-around ability is the yardstick, we win. In the air, we can fly from one handgrip to another more accurately (just watch a good trapeze artist's twisting somersaults). On the ground, where an ape waddles, a human can throw a dozen handsprings. Give us a cliff near water, and we'll learn to dive so precisely we can survive a hundred-foot, headfirst drop. Show us a snowy mountain, and we'll slide down on a couple of sticks at sixty miles an hour, launch into the air, flip a few times, land, and keep skiing.

We've invented these gravity feats not just because we're the only animal silly enough to think of them, but because we're the only animal with the talent to pull them off. When it comes to controlled falling, we have no peer.

The rise of extreme sports, and of ESPN's X Games in particular, has made stars out of those who push gravity's limits, including Mat Hoffman, shown here sailing above a vert ramp during competition. Hoffman, who has broken dozens of bones due to his exploits, has an attraction to gravity play that seems inborn. "When I was a kid," he told the author, "I jumped off my house with an umbrella, thinking I could fly. I was like six years old." *(Copyright © Red Bull / Bernhard Spöttel)*

CHAPTER 10

# Got Air?

These stunts should not be exposed to the young riding public.

—Cook Bros. BMX race promoters, referring to freestyle bike riding

ONE DAY IN 1993, as Ron Semiao sat slumped on his couch watching television, he began to notice something peculiar. While the show he was watching was a championship basketball playoff game, a lot of the commercials featured athletes doing nothing so traditional as dribbling a ball down a court. Instead, pierced kids with spiked hair were flipping through the air on skateboards, dropping from planes on skysurfing boards, or plunging from a tower attached to bungee cords. He began to count the ads that had some sort of extreme sport in them, and came up with fourteen.

This was no idle exercise for Semiao. As the chief executive in charge of programming for the television network ESPN2, he was charged with finding sports that went beyond those covered by his company's main network, ESPN. Suddenly, it seemed as if there was a new world right under his nose.

Semiao knew that for the last two decades attitudes in America had been changing; to many fans, professional sports had sunk to squabbles between rich owners and rich athletes, and the problem was that a generation of upcoming sports fans was being lost. Every kid did not want to grow up to play in the major leagues; to them, professional athletes seemed as distant as the rock stars whose bloated music they rebelled against with punk and grunge. The teenager who played sandlot ball and who would grow up to root for the teams on ESPN was becoming rare, if not extinct. In his place was a kid challenging gravity on a skateboard or snowboard or bike—and his sports were growing.

Semiao was aware that neighborhood competitions had expanded into regional events, that companies had sprung up to manufacture the equipment and sponsor athletes (some as young as fifteen), and that magazines were flourishing, covering the cult celebrities and filling their pages with advertising. There's an audience here, he thought, and we aren't reaching them.

Semiao, who started at ESPN in the financial department, knew how to follow the money. The key, he realized, was that instead of covering these new sports piecemeal, as ESPN2 had been doing with occasional shows on wakeboarding or BMX stunt riding, all of them could be brought together for one massive television event. That would enable him to attract major sponsors—including brands such as Nike and Mountain Dew—who were already exploiting extreme sports in their ads. In a flash of insight, Semiao had invented the Extreme Games.

What Semiao envisioned, though, was unlike anything the network had ever attempted. This was not a question of buying the rights to cover, say, a series of National Hockey League games; this was about creating a massive event from scratch. Yet the potential was irresistible: if it worked, it would be a historic merging of media and sport. It would be like owning the Olympics. It would also be the largest production ESPN had ever mounted.

After three years of planning, the network set about transforming Fort Adams State Park in Newport, Rhode Island, a site with no electricity, no running water, and no phone lines, into a massive outdoor playground for falling. "ESPN came in and built a city," said coordinating producer

Rich Feinberg. "Approximately six hundred personnel—between production, technical, and vendor staffs—lived in the city for a week."[1]

Invitations were sent to the leading extreme athletes worldwide to compete for $300,000 in prize money. Three hundred and fifty accepted, including virtually all of the top stars. The first games comprised twenty-seven events in ten sports: bungee jumping, barefoot water-ski jumping, kite skiing, windsurfing, skysurfing, bicycle stunt riding, mountain biking, street luge, skateboarding, and the Eco-Challenge, a cross-country race. A hundred and fifteen cameras—mounted on everything from an athlete's helmet to the bottom of a skateboard—caught the action and beamed fifty hours of it worldwide. When the numbers came in, ESPN knew they had scored a direct hit.

Nearly three-quarters of a million people watched each broadcast of the first Extreme Games. ESPN had planned to hold the games once every two years, but the 1995 event was so successful that another was planned for the following year, now renamed the X Games to better establish an exploitable brand name that would work worldwide. Almost immediately, a winter X Games was added; then a series of local competitions dubbed the X Trials, a demonstration event called the X Games' Xperience, and the first competition held outside the United States, the Asian X Games Qualifier.

In less than four years, X Games activities became a year-round, international activity for ESPN, a phenomenon with explosive growth unlike anything the network had ever seen. If there had been any doubt that this was the era of gravity recreation, at least for young men, the success of the broadcast put it to rest. "The X Games delivers the highest concentration of males 12–34 of any sporting event on television," says Jeff Ruhe, an ESPN senior vice president. "Take any one you like: Super Bowl, NHL Playoffs, Major League Baseball, auto racing, all the traditional sports that you see—more young people are watching this event than anything else out there."[2] At the 1997 X Games in San Diego, nearly a quarter of a million people came to see more than four hundred athletes compete in twenty-seven events. Thirty-seven hours of it were broadcast to 198 countries in twenty-one languages, reaching perhaps half of ESPN's 250-million viewers worldwide. By 1999, ESPN had added several more X events in countries all over the world and expanded the X Game franchise into concerts, music CDs, videotapes, and books. Unwilling to leave the future of sports to ESPN, in 1999 NBC mounted a rival extreme Olympics, dubbed the Gravity Games.

But is gravity play, as it's done in the X Games, really the future of sport? ESPN, with the backing of Disney, its corporate parent, is betting that it is. In a bid to actually own the next generation of sports fans, ESPN has now created the Junior X Games, an event with gravity action on smaller ramps. The target audience—and the athletes—are kids between the ages of ten and fourteen.

As a showcase for the newest and most rapidly growing sports, the X Games reveal just how fast falling has saturated popular recreation. None of the games' events could have existed even thirty years ago because they weren't yet invented, and all of them use gravity as their engine. Two of the sports—downhill inline skating and street luge—are variations on sliding, with athletes speeding downhill at up to sixty miles an hour. Sport climbing's artificial walls are built to ensure that even expert climbers will fall. In the rest of the events, athletes try to escape gravity altogether and for as long as possible. This is called "getting air." You get air by riding skates, skateboards, snowboards, bikes, and even bare feet off ramps, or by flying off a ski boat's wake on a wakeboard, or by dropping out of a plane at thirteen thousand feet attached to something called a skysurfing board. Once you've got air, you flaunt your freedom with the most complex gyrations you can manage.

On close examination, what extreme sports are all about is gravity—not about how gravity affects a ball or puck, but how it affects people. Random danger or terrific endurance isn't the point, which is why underwater caving and ultra-marathoning have had a tough time selling themselves as extreme sports, while wakeboarding (water skiing in a surfer's stance on a wide board), which barely passes the test with its short but spectacular falls from airborne flips, is an X Games event. To be extreme, you have to risk a fall, period. The more serious the risk, the more extreme the sport. Skiing becomes extreme when you spend more time in the air than on the mountain. Skydive from a cliff instead of a plane so there's no time to open a reserve chute, and you have skydiving's extreme cousin, BASE jumping. And while participation in the more dangerous sports will never rise to that of, say, golf, those that offer the thrill of being airborne in safer ways are quickly becoming mainstream.

## Mat "the Condor" Hoffman

In 1998, the X Games were held at Mariner's Point, a small peninsula that juts out into San Diego's Mission Bay. Normally, this is a peaceful city playground that is half park, half beach, but it was hard to see that once the X Games had taken over. The air was fresh and cool, even though it was June, and a saltwater breeze blew in from the ocean. After parking and being checked for my press credentials at the entrance, I followed thick TV cables that snaked in bundles over the grass to the center of the park, where cement benches and picnic tables squatted among the huge structures built for the games. Here the snowboard jump awaited polar cannons that would appear later in the week and spend sixteen hours blowing a hundred and thirty tons of snow on it. Imagine a frosted football field tilted up on one end to the height of a ten-story building, and you have a good picture of the snowboard ramp. Nearby, a fifty-foot climbing cliff leaned out at a steep angle. Everywhere there were huge towers of black scaffolding topped with white-tented fortresses, each stuffed with cameras and sound equipment. Truck-sized video screens on thirty-foot pedestals seemed to be around every corner.

From the park I made my way to where I was to meet bicycle stunt rider Mat Hoffman. I recognized him immediately from the photos and television interviews I had seen. He seemed taller and leaner in person, with sharp features and penetrating eyes softened by a mop of curly dark hair. Before I shook his hand, I noticed he was hobbling, with a hinged cast on his knee. I was not surprised. As a stunt rider—in Mat's case, the world's best—injury is not at all unusual, and neither of us mentioned his cast.

Mat rides a BMX bike. These are undersized bikes—seemingly kid-sized—with flat handlebars, small, twenty-inch wheels, and raised seats, and to withstand the pounding they regularly take, they are built like tanks. BMX is the abbreviation for "bicycle motocross"; the bikes evolved from the old Schwinn Stingrays into models that boys could use to imitate the motorcycle racers they saw jumping the hilly dirt tracks of the early 1970s.

In the gravity culture of the late seventies, it didn't take long for BMX riders to realize that if a skateboarder could launch himself into the air from the side of a swimming pool, a bike rider could too—and with a lot more speed and height. By the early 1980s, some BMX riders were breaking into boarded-up skateparks to see what they could do. Stunt riding then grew as an outlaw sport, certainly because parents worried about bashed-in skulls but also because the profitable BMX racing circuit was rapidly losing kids who wanted to ride what came to be known as "freestyle." In tirades reminiscent of the criticism heaped on gravity activities of the last two centuries, the BMX establishment tried to squelch freestyle with dire warnings. One race promoter released a statement saying, "Cook Bros. does not condone stunt riding with BMX equipment. Our effort is in racing, not 'circus acts.' Furthermore, we feel these stunts should not be exposed to the young riding public."[3] Statements like this backfired of course and became a siren call to any self-respecting BMX rider. By the mid-1980s, stunt competitions and sales of heavy-duty bikes made for the purpose skyrocketed.

It's not hard to understand why parents would be terrified of BMX freestyle. On the black asphalt tarmac the size of a half-dozen basketball courts that had been spread over the beach here at the X Games, I got a firsthand look at the risks. Boxed in with high-rise bleachers and packed with dozens of black plywood ramps in all shapes and sizes, this was where skateboarders, inline skaters, and the BMX riders would compete in the street competition, or just *street* for short. While waiting for Mat, I watched the street riders warm up. In their head-to-toe leathers, heavily gloved hands, and with heads replaced by bubble helmets, they looked like astronauts who had awakened to find kids' bikes as their only means of launching themselves off this alien landscape. And it looked as if launching themselves into the stratosphere was exactly their goal.

They flew off the ramps up to twenty feet in the air. As I was standing next to something called a spine, which is two back-to-back six-foot ramps that meet in a point, a rider shot up and off. Five feet in the air, he seemed to float for a second, slowly turning his bike nose-down to connect with the ramp on the other side. His front wheel hit with a bass-drum boom, and he let out an exhaled "huh!" before pedaling off with a savage fury.

With every jump a rider risks coming down on forty pounds of protruding bars and pegs. As he sails over a jump box, a cube with ramps on all sides, he might get off the bike in midair, stretching out like Superman and hanging on only by the handlebars. Or he may give the bike a kick, sending its tail spinning around beneath him as he flies straight, holding the bars, and hopping onto the seat as it comes back around under him; this is called a tail whip. He might instead hunch up onto the handlebars and whirl around as if on a turntable attached to the front of his airborne bike, dropping back into the saddle just in time to land. Or he'll do both at once, spinning the bike one way while he spins himself the other way. Then there's a trick called a nothing. In this, he lets go of the bike completely while flying; it sails on, leaving him behind until he grabs the seat and yanks himself back on.

A rider doesn't have to spin to find trouble. I watched one rider try to clear a handrail that ran along the side of the jump box. Something went wrong; he shoved his bike out of the way but he came down on the rail squarely on his crotch. For an agonizing moment he turned slowly over to hang upside down from the rail. Then he lowered himself to the deck. In a few minutes he was up and on his bike.

Flips are more dangerous still. In a flair—a back flip with a half-twist—you're off a ramp, ten feet in the air, backward, upside-down, crooked, spinning sideways, and coming down fast. If it works, you land on the same ramp rolling forward. If it doesn't, there's no telling. Maybe you land on your head, feet tangled in your machine. Maybe you smash onto the bike, the handlebar jammed into your stomach.

In one of his more spectacular crashes, which was captured on video, Mat launches off a six-foot ramp, shoots up twenty feet, and begins a double back flip. After the first rotation he starts to tilt, no longer perpendicular to the ground. By the time he is halfway through the second flip, he's spinning sideways. He's upside down when he bails, trying to kick the bike away, but it's too late. He smashes into the deck on his face and chest, his back arched, his feet still on the pedals. As he collapses, the bike falls on his legs.

Known tricks that misfire aren't the only source of crashes, because riders at this level have to be willing to invent tricks, and there's no guarantee they'll work. Later, watching the competition, I saw Dave Volker try a trick that was, as nearly as I could tell, impossible. He flew off a ramp and slammed his bike's tires into a wall, as if he were going to ride it straight up. Then he jerked hard on the handlebars, trying to will his machine into a back flip from this dead stop. He made it about halfway. He hit the pavement flat on his back.

As dangerous as street competition is, the riding Mat is best known for is more dangerous still: this is vert riding. Vert ramps (short for vertical) evolved from skateboarders' passion for riding both swimming pools and the huge cement pipes they would find at construction sites. The vert ramp is a cross between the two. It looks roughly like a section of an enormous pipe with its top cut off: a gigantic U. The two curved parts that face each other are about twelve feet high and rise straight up to flat platforms on either side. The bottom, rather than being curved like a pipe, is flat, about fifty feet across.

On this contraption, a rider whips from one side to the other, flying straight up a dozen feet or more above the ramp's lip and nearly thirty feet above the ground. Airborne, he will spin or flip, reenter the ramp, shoot across the bottom and up the other side, and in less than two seconds be high in the air again, spinning or flipping. The action is constant; there is no time to think, only time to act, and the best riders link one astounding trick after another.

Nicknamed "the Condor," Mat can fly above the ramp and spin himself and his bike two and a half times before his wheels touch down; then, rocketing into the air on the other side, he might top things off with a *double* tail whip. With perhaps forty high-flying tricks packed into a ninety-second run, and nowhere to go but straight down when you make a mistake, bike vert ensures pain. "If you're going to enter a pro vert competition," Mat has said, "don't do it if you expect to walk away healthy."[4]

When people list Mat's accomplishments, they also list his injuries. Mat won the vert ramp World Championships ten years in a row. Over that time he's had sixteen surgeries, half of them major reconstructions. His spleen has been removed. His shoulder has been rebuilt four times. In a televised interview, Mat pulls back his shirt sleeve to show the four-inch scar. "I've completely torn my rotator cup off," he says. "I don't have one any more."[5] Mat is twenty-five-years old. At this point in his career he seems almost oblivious to pain. In practice at the 1996 X Games he broke three bones in his foot but chose to compete anyway, a piece of wood and tape holding his foot together inside his shoe. He took first place.

Mat and I sat down inside a refreshment tent near the street course, and I waited for a moment while a mom encouraged her shy twelve-year-old to ask Mat for an autograph. Mat was gracious and chatted at length with the wide-eyed boy.

Mat has about him a relaxed confidence. Although his company, Hoffman Promotions (one of seven corporations he owns), is in charge of managing the stunt bike event here at the X Games, he seemed to have all the time in the world to talk with me. Dressed in a plain brown T-shirt, shorts, and sneakers, Mat was utterly without guile or pretense, and it was hard to keep in mind that he is a successful businessman whose concerns include a bike manufacturing company, a clothing company, and a retail mail-order business, all of which bring in several million dollars annually.

Trying to capture Mat on paper is a problem. He talked in constant revisions, changing the beginning of a sentence two or three times before going on to add the end, which he would also revise a time or two. Sometimes he worked through three or four false starts before his real sentence began. It seemed like a verbal reflection of what he does in the air: acutely aware at every millisecond of his flight, he was constantly adjusting, refining, doing everything possible to ensure a safe landing—precision as a tic. He didn't strike me as someone with a death wish, and indeed, he said he doesn't have one.

I asked how the thrill of what he does could possibly be worth the injuries he's had. "I've always looked at it like if you want to experience all the pleasures and successes life has to offer, you have to be willing to accept all the pain and failure it has to offer," he said. "You can't have one without the other. And I like challenging the body, too. It's a pretty amazing instrument—it keeps healing itself. Plus, with modern medical science, I may as well take advantage of it."[6]

This hardly seemed like a sensible explanation. For a moment, I wondered if Mat's desire for flight was pathological.

"When I was a kid," he told me, "I jumped off my house with an umbrella, thinking I could fly. I was like six years old."

Did you hurt yourself?

"Yeah, I got hurt. Then I jumped off my barn. I tried to jump on to my horse, but it moved."

I wondered what his parents thought about this.

"They were funny. I remember one time I built a hang glider out of two-by-fours and a sheet, and I was going to jump off the top of my slide. I was all stoked. I was up there and I was arguing with my brother about who was going to fly first. So we decided we were both going to do it at the same time. And we called down to our parents, 'Come check it out!' They came out just shaking their heads, but they didn't say anything. Just let us do it. Travis broke his finger. Had to go to the hospital."

How old were you when you did this?

"I don't know. Maybe like eight."

Mat's lust for air hasn't changed much since. Not satisfied with the height he got from the standard twelve-foot ramp, he built one that was twenty-one-feet tall and had a friend with a motorcycle tow him in on his bike at sixty miles an hour. He shot up nearly fifty feet off the ground. In 1997, he rode his bike off a cliff in Norway. The cliff was more than three thousand feet high. After turning a couple of back flips, he kicked the bike away and pulled the rip cord on his parachute.

Mat admits that he doesn't have much control over his craving. "When I did that jump with my bike I thought, 'How am I going to top this?' Now it's just going to start getting more and more life-threatening. And that's how you live short. So that's what I was a little worried about. Because when I get these ideas in my head, I'm just along for the ride. My will takes over. And my determination takes over. So I was hoping that I wouldn't get some crazy idea that was even crazier than that. And luckily, I didn't."

He credits his will and determination—and his fear—with protecting him so far. "That jump was the scariest thing I've ever done. Fear gives you that extra energy that gets 110 percent out of you. If you didn't have fear, you wouldn't have that intense respect and intense concentration to make sure you don't mess up."

The most striking thing about our conversation was the matter-of-fact way in which Mat described what he does. He talked about it the way you or I would talk about the perils of romance—that in spite of the pain, living without it wouldn't be much of a life. I began to see his point. I found myself believing that this was all completely reasonable; that it was the nature of life, not the quirk of this particular man.

It was also not the pressure of the crowd. Unlike Evel Knievel (who, I discovered, has become a friend of Mat's), Mat doesn't have to up the ante with each performance to attract an audience. There were no spectators at his cliff jump; he does these things for himself.

Yet he doesn't do them impulsively. They hardly provide the kind of instant gratification you might think would go with so strong an urge. Of the jump off the cliff he said, "It wasn't just like a 'wahoo!' thing where I came up with the idea and I'm like 'all right, let's pack up and go to Norway.' " In fact, he spent seven years preparing, first learning traditional skydiving, and then the extra skills he would need for BASE jumping—free-falling from a fixed object like a cliff. "You need to know how to track. You need to know how to keep yourself in control, get away from obstacles, canopy control, and how to land. There are so many things you need to learn. Because in BASE jumping there is no such thing as an injury."

I pressed Mat to describe the actual feeling of his fall, trying to get to the heart of the sensation itself, but it seemed elusive to him. His comments about it were brief and interwoven with the physical description of what he did. "It was just so intense, it was beautiful," he said at one point. Later, "You feel like you're in danger. You get that rush of adrenaline. But the object is to be in complete control and to pull it off. And when that happens, you just feel accomplished."

Of free-falling from a plane, he said, "You never really feel the way you do in a roller coaster, where your stomach drops. It's not that kind of feeling because you always have some sort of pressure on you. You're falling so fast that the wind is so intensely around you that you feel like you're on this bubble. So it's not really like you're 'wahh!' falling. It's not like a vacuum. You fall and then you can't fall any faster, because the wind is holding you back." BASE jumping, he said, is different, "because you jump off and it takes about five or six seconds before that wind starts really hitting you and you can control your body and start flying. You actually use your arms and your body like rudders in a plane."

I asked what was enjoyable about it. "I guess it's just that you're not supposed to be there." He said this twice. "You weren't born with wings on your back so you're definitely pushing the limits if you're there. So it's kind of cool to cross that line."

I asked if he thought there would come a time when the risk would be too much.

He thought about that for a moment. "That's the road I fear to pass," he said finally, "when my mentality just cannot relate with my physicality. My head's always going to be this way. It's the way I think, the way I am. But someday it won't correlate with my body. I definitely dread that day. I'll just have to figure out a creative outlet that isn't so physically challenging. I won't know what that is until the time comes, because as long as I can use this as my outlet, I'm not going to be desperate enough to find something else."

## "You Don't Want to Do That"

As I walked down to the beach, I caught a glimpse of barefoot water skiers showing how falling had worked its way into their sport. Beyond a white mesh fence, a forty-foot boom swung far out over the water to follow the whizzing barefooters flying from ramps. Attached was a high-speed camera that expanded two-tenths of a second into a full minute: its job was to catch in exquisite detail all the air a barefooter has gotten. An excited announcer called out the tricks, blaring through a row of black outdoor speakers the size of filing cabinets.

Water skiing has been around since the 1920s, and barefoot skiing since 1947, but it wasn't until the late 1960s that anyone considered jumping off a ski ramp barefoot. *Considered* is probably the wrong word, though, because it implies a sensible evaluation of the hazards. Barefooting itself is dangerous enough, because while the typical speed on a water ski is around thirty miles an hour, barefooters must ride at close to forty-five miles an hour just to stay on top of the water. At that speed a fall can twist your arm into a pretzel. To jump means to hit a solid plywood ramp with your bare feet at freeway speed, fly high over the water, and then somehow hit the surface without breaking your neck. Picture jumping from a high dive with the water below looking like the asphalt under a speeding motorcycle, and you get some idea of the challenge. All we know about who did it first was that the practice arose among a group of Australian barefooters known as a pretty macho crowd. As far as details such as dates and names go, no one remembers. My guess is that it was the result of a drunken bet.

By 1978 tournaments were being held that included barefoot jumping, although the technique was still being refined. William Farrell was getting good results with what he called "bum jumping"—sliding off the ramp on his butt—a posture that a lot of jumpers used until it was banned in 1988. Everyone else left the ramp on their feet and also tried to hit the water on them, attempting to stay upright through their flight. The distance record climbed gradually until it hovered around seventy-six feet. Then in 1989, Mike Seipel lost control during a jump and, still holding on to the ski rope, stretched out like Superman, which although far more dangerous than standard practice (picture a belly flop at forty-four miles an hour), seemed to get him farther across the water. The "inverted jump," as it was called, remained just an option until 1991, when Jon Kretchman proved its superiority by beating the old world record by almost ten feet. By 1996, everyone was using the inverted style, and the world record climbed to more than ninety-two feet.

As I arrived at the shore, an X Games official was introducing three barefooters: Rael Nurick from South Africa, Brett Sands from Australia, and Lane Bowers from the United States. They were there to answer questions. Out in the bay a skier began his pass. He leaned back, with a tremendous rooster tail spraying from his feet. He suddenly dropped to his back and skidded along the water, spinning around like a break dancer, one of the tricks barefooters do as part of this event. In a moment he was up and racing toward the ramp, which at only eighteen inches high appeared too small to give him much of a lift, so what happened next seemed impossible. He hit the ramp and exploded into the air, flying twenty feet above the water and looking like a trapeze artist whose rope was attached to a rocket. As he began to drop, he pulled himself into a crouching position. He plunged into the water and disappeared in a blast of spray, utterly gone. Then he somehow emerged from the foam, stood up to ride the surface again, and waved to the crowd.

Lane Bowers explained what was different about the jumps they were doing compared with their usual competitions. "This event is set up especially for the X Games. In our traditional skiing, this would be two separate events. You'd have a tricking event without a ramp, and then you'd have a jumping event, where you only line up and jump over the ramp. For the X Games, to make it more spectacular, they've combined the two. All this tricking that's being done before the ramp is completely disorienting the skier. You have to be able to turn the tricking on and then turn it off, stop your head from spinning, focus enough to go over that jump and concentrate, because you're spinning one way, you're spinning the other way. That's what makes this event so extreme."[7]

I asked the skiers if one hundred feet was a record that people were aiming to break. They agreed it was. Lane said, "They might have to add more speed or even a couple of inches to the jump, extend it even a couple of inches, add another mile an hour or two." (The official speed is precisely forty-four miles an hour; for the X Games, it is forty-four and a half.) "But the people who regulate our sport are very restricted by safety concerns, so any increase in speed is tough to get out of them. We'd probably all take it at least forty-five miles per hour, maybe forty-six if they'd let us. Definitely get us going farther."

I asked Lane if they had jumped at that speed.

"Yeah, you get a bad pull from somebody sometime, they're driving you funny, and they end up pulling you fast, but you're not allowed to in competition. And also for the X Games there's an additional half mile an hour which isn't allowed in our traditional skiing. So we are going faster than normal." This means they can't break a world record because the speed is beyond what's allowed. The skiers said they could tell within a half mile per hour how fast they're going.

On the next run, we saw a skier fall just in front of the ramp. Someone asked what he did wrong. "Coming in to the jump, what happens is, you start thinking about your raise, and you start leaning forward," Lane explained. "In barefooting, you have to keep this part of your foot out of the water"—he pointed to the part forward of the ball of his foot. "If the water comes up to here, or here," he said, moving his finger two fractions of an inch toward his toes, "you're done. And that's exactly what happened. You start thinking about it, rocking forward, your feet go under just a little bit, and you're done. That fast. It happens real fast."

Falling by itself can be dangerous enough—if your arm or leg catches wrong, you can easily tear a muscle—but falling in front of the ramp is much worse. Skiers wear helmets, but, as Lane pointed out, at these speeds, that's not enough. "If he had fallen two or three feet farther, he definitely would have been 911. Because you see how his head's moving so fast forward, you're going forty-four, when you fall you probably accelerate to fifty or above, and you're hitting a stationary object that's coming up. If you fall within five, six feet of the ramp, you could die. It's pretty serious." Lane laughed. "You don't want to do that."

As a sport that grew outside the mainstream for years because it was considered an outlaw activity by many adults, skateboarding culture developed its own ethics and dialect, especially concerning tricks. Here Andy Macdonald performs a "benihana" high above a vert ramp. *(Photo by Grant Brittain, copyright © Andy Mac Productions)*

CHAPTER 11

# Falling for All

It starts with a bunch of kids just raising hell. —ARLO EISENBERG, INLINE SKATER, ON EXTREME SPORTS CULTURE

UP UNTIL THE 1960s, there had been little in common between daredevils and their spectators, who only got an occasional dose of the feeling itself from a slide down a slope or from the packaged falling of an amusement ride. Today, the gap has been closed. The teens who in the 1970s took cheap, common vehicles like skateboards, skates, and bikes into the air erased the line between daredevil and dilettante. Every weekend BMXer in the X Games bleachers knows he's in the same game as Mat Hoffman—Mat just flies higher.

The X Games athletes feel this in their bones. Although I've used the term *athletes* and called what they do *sports*, that's only for a lack of better terms. Skateboarders, inline skaters, stunt bike riders, and many of the others don't think of what they do as sports. Their feelings about it run much deeper. "This is the difference between a sport and a lifestyle," Arlo Eisenberg, perhaps the best-known aggressive inline skater, has said in a television interview. "I don't like it being reduced to a science. Because what we do is very unscientific. It's artistic. It's expressionistic. It's about individuals. We're challenging this whole conventional, authoritarian structure,"[1] Arlo said.

Seen from the perspective of this generation, team sports, with their reenactments of battling armies, seem like an anachronism from the Middle Ages. Which means that ESPN, in working with the athletes, has its work cut out for it. The network has sometimes tried a little too hard to fit these kids and their sports into traditional sports coverage, while the athletes are intent on bursting seams. When one was asked to pose for the camera, take off his baseball cap, and hold up a title card, he refused, saying simply, "Too many rules, dude." To many skateboarders, the idea of fierce competition against another skateboarder is laughable—so is quantified judging—because skateboarding isn't a game like tennis or golf, where someone wins and someone loses. Gravity sports are tribal, born from kids hanging out together. To a large degree, the athletes at the X Games are still amateurs.

One young ESPN producer, Sean Pamphilon, explained the difference to me. "You interview a player in the National Football League and you have to go through publicists, you have to go through media relations, and often times because of their schedule and time constraints, you're given like ten to fifteen minutes with them. These guys you hang out with. I leave with them the night before, have dinner, or go out or something, and the next day I pretty much have the whole day with them. I think a lot of young people are gravitating toward these sports because of the simple fact that these guys seem more accessible when you hear about them, when you read about them, when you see them. And they are fan-friendly; they're people-friendly. I mean, I couldn't imagine seeing a stunt cyclist having an agent."[2]

Yet ESPN knows that these athletes are becoming celebrities, whether they want to be or not, and treat them as such. ESPN's brightest spotlight, and most deferential treatment, is reserved for Tony Hawk, who was already acknowledged to be the best and most famous skateboarder in the world long before the X Games.

To say that Tony is the most well-known skateboarder in history is a pathetic understatement. He is to skateboarding as Einstein is to physics, or Freud is to psychology. He has been around so long and is so good that many skateboarding fans are sick of him. But everyone knows who he is. Tony, blond and clean shaven, his lanky six-foot-two-inch body sticking out of baggy shorts and a T-shirt, is now in his early thirties and has been a professional skateboarder for nearly two-thirds of his life.

Tony first stepped on a skateboard at age nine and learned to become airborne by flying from the lip of empty swimming pools. By the time he was a senior in high school he was earning a six-figure income. In most of the years since, it has been nearly impossible to dislodge him from the top spot of any competition he's entered. He has won the National Skateboard Association series eleven years in a row. Dozens of the vert ramp tricks skateboarders now regard as standard were invented by Tony. Among the often repeated things said about him is that when he enters a competition, there's going to be a hell of a fight for second place.

There have been upsets, however; at the 1996 X Games Andy Macdonald bested Tony in vert. But you don't have to watch Tony long to see that, as with nearly all X Game athletes, rivalry isn't what motivates him. Tony applauds when another skater lands a good trick, not just in competition, but any time. During my visit to the 1998 X Games, he told me that although he qualified for street, he decided not to enter. "I just feel like I want to have a real street skater representing our sport out there," he said. "And the next guy down that qualified was Alan Petersen. He's a street skater, so I just thought, he's going to show what street skating is more than I am."[3] In the double vert competition that year, where two riders team up to work the ramp in a tightly choreographed routine, Tony rode with Andy, the man who beat him two years earlier.

When during practice Tony called out to nobody in particular to ask where his four-year-old son had run off to, he got star handling—in spite of the fact that Tony is an utterly unpretentious, regular guy. A male producer in shorts, sunglasses, and a headset instantly sent an assistant to track Riley down. She found him on his way up the stairs to the vert ramp. The producer then spoke to the four-year-old as if he was a Hollywood drinking buddy. "Riley, you wanna come with us, bro?" he shouted up. Riley looked confused. "You wanna hang here?" Riley reluctantly descended. The producer then took his assistant aside. "Watch Riley," he ordered. "You're a ten-minute baby-sitter."

Having invented the Olympics of gravity sports, ESPN is now struggling with the antiauthoritarian values of their athletes. It's not easy, especially when the values spread to ESPN's own producers. To its credit, ESPN has attempted to give voice to these values; inspired by what he saw in these athletes, Sean Pamphilon created a segment of commentary that aired during the 1998 X Games—although not without a disclaimer by the Games' host, Chris Fowler. "While we don't necessarily endorse all of his thoughts," Fowler noted by way of introduction, "he provides a word of caution to the athletes we've seen over the last week."[4] In the segment, Sean speaks to the camera:

> I have this recurring nightmare. I'm trying to set up an interview with a sixteen-year-old inline skater who insists that I go through his people. . . . I'm part of a lost generation of sports fans. . . . I just want to sit in my Lazy Boy and not be disgusted. Let us root for you. Dennis McCoy, you're my Don Mattingly. Tony Hawk is no Michael Jordan, he's Joe Montana, the chill genius who couldn't explain how or why. The reason why we like you is simple. You don't charge us for autographs. You don't have vanity plates on your cars, and you haven't spoken in the third person while analyzing your global impact and contribution to world peace. Look in the mirror. Realize and recognize. Hey, flavor of the month, savor it while you can, because the day you become a personality, you become a commodity. You lose touch with reality the first time you buy into your celebrity. I want to be ten again. I want to get goose bumps. I want to root for somebody. . . . I'm announcing my free agency as a fan and I want you guys to sign me.[5]

ESPN executives will continue to wrestle with how to best work with these gravity antiheroes because they have little choice. Falling, in all its guises, is the perfect modern sport. It is democratic, because it pits all humans equally against the inanimate threat of gravity, and yet individual, because each athlete meets the threat alone. It appeals to the modern urge for immediate gratification, because gravity is always there to be challenged. And it comes in TV-sized bites, its three-act dramatic structure of launching, falling, and landing fitting neatly into a few seconds, perfectly suited for the quick-cut MTV style of voyeurism that has permeated popular entertainment.

Another reason ESPN must struggle is because competition is breathing down its neck: as influential as the X Games are, there's no reason they can't lose athletes to shows more attuned to a new attitude. They can also lose their staff. Sean Pamphilon has already left the network. He now works for NBC's Gravity Games, which from the beginning promised to focus as much on the athletes' lifestyles as competition.

## Vert

At the twin vert ramps where the skateboarders were practicing, I took a seat in the bleachers to watch Andy Macdonald, winner of the 1996 vert competition, show what a human can do while sailing through the air on a wheeled piece of wood.

If vert bike is the prize fighting of extreme sport, vert skateboarding is its ballet. Without the heavy machinery and flight suits of the BMXers, skateboarders' fluid movements show through their fluttering T-shirts and shorts. The terror of watching is gone—vert skateboarders are rarely seriously hurt; a botched spin usually sends them sliding safely down the ramp on their kneepads. In the air, the choreography includes the board, which is often twirled on fingertips like a dance partner that rejoins his mate for a final flourish.

Andy began by standing on the top of the ramp. He set the tail of his skateboard on the edge, pinning it down with his back foot so it hung out in space. When he was ready to drop in, he stepped on the front with his lead foot, pushing the nose straight down the vertical wall.

As he reached the basin and the board came under his weight, he crouched, rose, then crouched again, pushing smoothly against the ramp for speed. At the far wall, he flew six feet above the lip. With one hand he held the board to his feet; he threw the other into the air and began to pirouette. He spun one and a half times and slid into the ramp.

Suddenly, as he came down from one of his flights, there was a horrible accident. He came in all wrong, short of the ramp. The board smacked onto the ramp's edge between the front and back wheels and stopped dead. Andy leaned into the gulf. By some miracle he snapped down on the front of the board just in time to unhook the back wheels before he tumbled, then shot down the ramp. I looked around the bleachers with a did-you-see-that look, but nobody was alarmed. Andy had just pulled off a nicely executed trick called a Disaster.

Among skateboarders at this level, a Disaster is routine, along with a lexicon of moves as formal and extensive as any in classical dance. There is the Cabbalerial, Gay Twist, McTwist, Half Cab, Invert, and spins known as 180s, 360s, even 720s—the number of degrees a rider turns in the air before landing. These tricks (and many more) are augmented with how the rider approaches the trick (a backward approach is called Fakie, for example), how he grabs the board (Nosegrab, Tailgrab, and others), whether he is on the board in his usual stance or a "switch stance," and whether or not he spins the board underneath him with his hands. Instead of spinning in the air, skateboarders often slide along the edge of the ramp; this is called grinding, and this maneuver too has several variations, depending on what part of the board or axle is slid on: Boardslide, Noseslide, Blunt, Nosegrind, Crooked Grind, and many more. Grinds seem impossibly unstable. In a Crooked Grind, for example, the back wheels of the board roll on the flat deck of the ramp's top while the front wheels straddle the lip, sliding along on the axle, one wheel hanging off the edge. Compound tricks get their own names and include the Miller Flip, Switch Back-Flip Mute Grab, Nollie Frontside Hurricane to Fakie, Stalefish Nosebone, Judo Air, One-Foot-Tuck-Knee Invert, Fingerflip Varial Invert, Frontside Boneless Fakie, and Backside Disaster Revert.



When Andy broke for lunch, he and Tony Hawk headed to the food tent, and I fell in behind. Dressed in a T-shirt and sneakers, Andy seemed more like a graduate student who never left the dorms than a world-class athlete.

I asked him if it was being airborne that had led him to skateboarding and to vert. "Yeah. Totally. Vert was the form that attracted me most because of the things people did in the air." I mentioned that I found it incredible that on many tricks skateboarders can keep the board on their feet without using their hands. "Just ollies on the vert ramp, you mean," he smiled.[6]

Yes, just ollies. What Andy is talking about is a technique that along with pool-riding, revolutionized skateboarding two decades ago. Almost from the sport's beginning, skateboarders learned to make kick turns by pushing on the back of the board, lifting the front wheels off the ground, and spinning on the back wheels. They also knew how to roll off curbs and land on their board as they fell. What they didn't know how to do—what no one considered possible—was to keep the board on their feet while they went *up*. In the 1970s, a thirteen-year-old in Florida named Alan "Ollie" Gelfand began to amaze his friends with a trick he had invented. Alan could get his board to jump up in the air by snapping the tail down; by lifting his legs at precisely the same time, he and his board would fly. When pro skateboarder Stacy Peralta saw the trick in 1977, he brought Alan to California where his technique was captured for *Skateboarder* magazine. The ollie has since spread worldwide.

The ollie is a breathtaking act of coordination. On the street, it looks like this: first, the skateboarder sets his lead foot well back from the front of the board. Then he kicks down with his back foot on the tail to pop the board into the air. He uses the edge of his lead foot to simultaneously drag the board up and to guide it into flight. This takes exquisite delicacy; if he applies too much pressure, he'll stop the board from rising; too little, and the board will shoot out from under him. Once the board is flying, his lead foot bears down, bringing the board parallel to the ground. Done well, the rider will suddenly find himself squatting on a skateboard perhaps four feet off the ground, as if on a magic carpet. Finally, both feet press down for the landing—a landing that might be on the street, a curb, a flower box, or even against a vertical wall. Using the ollie, a good skateboarder, rolling on flat cement, can jump up and over a row of waist-high trash cans. Confronted by a downward set of stairs, he can leap to the bottom and skate away, or he can ollie up to the handrail, land on it sideways, his board across it like a see-saw, slide to the end, drop off, and hit the ground rolling. A better skateboarder can do this same trick but slide down the rail on just the front tip of the board.

On the vert ramp, an ollie seems even more impossible because it happens so fast that you can't see what's going on. As a skateboarder shoots up the vertical wall, he punches down on the back of the board just before he leaves the ramp. He is now in the air, the board still on his feet. He arches high over the lip, flying in a half-circle trajectory, his body stretching out horizontally to the ground. He turns, rotating to face the fall, ready to bellyflop if things don't go right. And there's the board, still on his feet, as it connects with the ramp and he rolls down the wall.

I asked Andy what's next. "I think that inverted aerials are the next step," he said.

By inverted, you mean a back flip?

"Yeah. It's been done on bikes and it's been done on inline skates. And just like last year I did the first one on a skateboard. Just over a jump box. More like a side aerial, you know, a side flip, because you're riding sideways, but the hardest thing is to grab your board, because you're not strapped on. On a bike you hold your handlebars, on inline skates you don't have to grab anything, because you're strapped in. So on a skateboard, it's ride up, shoot the board up into your hand, and then do your flip."

I asked if he would try a back flip this year in the X Games.

"I hope so. A lot of it has to do with speed. If I get the speed out there on the street course, I'm going to maybe try one. But I see skateboarding going that direction. I think people are going to start trying it off vert. Just being able to go straight up, back flip, and coming in backwards."

Have you tried it on vert?

"Nah. Too scared. Way too scared."

As it turned out, Andy did not attempt a back flip that year.

## Boards

The uncanny skill that the best skateboarders display is every bit the equal of traditional gymnastics, with one important exception: great strength isn't needed. What extreme sports have discovered, and what has made them so accessible to so many people, is that with the right tool nearly anyone can fly. That tool is the board: the board is the wheel of gravity play.

The essence of a board is that it gives us a stable place to stand against the water, slope, or air we choose to fly from. Surfing was the first to show the board's promise, and most gravity sports have simply applied it to a new surface, fine-tuning its design for each particular use. Nature's ramp, an ocean wave, inspired skateboarders to find the shape in concrete; snowboarders, not content with natural slopes, now build vert ramps out of snow. The wake behind a ski boat, which was long used as a ramp by water skiers, didn't see its full potential until wakeboards were invented. Slowing down the boat a bit to create a larger wake, and with more surface to push against than a single water ski, wakeboarders can now somersault eight feet over the water.

When a French skydiver named Joel Cruciani strapped a surfboard to his feet in 1987, he found he could ride the enormous cushion of air beneath him as if it were an endlessly upturning wave—and skysurfing was born. With refined boards, skysurfers began to shoot back and forth across the heavens for hundreds of yards. Almost immediately they also found they could dive headfirst and spin the board against the air like a propeller, a trick now called a helicopter.

Besides the board and the surface it is applied to, there is one more ingredient that makes the acrobatics of extreme sports possible, and that is the sideways stance, also a gift from surfing. It is a position that maximizes our already acute ability to balance. Putting your feet sideways on a board lets them work as levers to control side-to-side motion much more efficiently than if they were pointed straight ahead. Legs spread wide gives you front-to-back stability, controlled by the fulcrum of your waist and the shock absorbers of your knees. Twisting your torso, you can use your arms as counterweights wherever they're needed, anywhere in 360 degrees of rotation. It is a powerful combination.

Look at the difference between a skier and a snowboarder's body position, and you can see why skis work better on a slope and a board works better in the air. Facing forward, most of a skier's flexibility is concentrated along the front of his body, his knees working like the independent suspension of a car, absorbing the bounce of each mogul separately. He can bring his skis together or separate them to handle the changing terrain. There's no difference between his side-to-side flexibility, so threats against balance from either direction can be handled equally. On a slope, a snowboarder has more side-to-side flexibility than a skier, but bends at the waist to lean one way and at the knees to lean the other—a more complex and asymmetrical set of movements. With his feet locked to the board, a snowboarder's ability to conform to bumpy terrain is much tougher.

In the air, though, everything changes. Suddenly, skiers have a couple of loose airfoils on their feet, each of which wants to go its own way with the pressure of the wind. If they're tangled or spread on impact, the skier gets a mouthful of snow; it takes strength and skill to keep skis together for a smooth landing. For the snowboarder, the triangle of his two legs locked to his board works like a single, powerful limb with a broad, stable foot. The board may destabilize in the air, but the snowboarder has all the muscles of both legs and torso working in unison to bring it under him before landing, and when he hits he's got just one large platform to worry about. Flips and spins are far easier, landings more reliable. Even weekend snowboarders can manage a short arc of air above the lip of a snow vert ramp, while true vert tricks on skis—where you reenter the same ramp you left from—are best left to the experts.



While getting air can be achieved on bikes and inline skates, board sports on water or snow offer the easiest way up and the most forgiving landing. It's easy to see why surfing, wakeboarding, and snowboarding are the gravity sports grabbing the majority of weekend athletes.

## X-Rated

Rock climbers are an anomaly at the X Games; their traditions didn't come from surf culture or its descendents but from the much older history of mountaineering. Yet climbing meets the criteria for an extreme sport: you can fall. To popularize the sport, a way was needed to make climbing convenient and controlled, and so like the vert ramps invented by skateboarders, climbers invented the artificial cliff.

The one at the 1998 X Games looked like a gigantic stone wedge that had dropped from the sky and been driven into the earth. The business side of the fifty-foot wall leaned slightly outward and then arched into a serious overhang. Fixed to the wall were artificial climbing holds: fist-sized plastic stones in dozens of different shapes that were bolted to the wall according to the design of the route setter—in this case, Mike Pont.

I watched Mike for a while as he yelled his conversation with climbers on the wall who were setting the holds. Like most accomplished rock climbers, Mike is small, thin, and well muscled. He's worked as a route setter for climbing gyms, contributed to a rock-climbing book, and has designed rock-hold sets—these are bolt-on holds an amateur can use at home to build his own artificial wall. He's also on the board of directors of the American Sport Climbers Federation, the governing body for competitive climbing in the United States. That puts Mike firmly on one side of yet another divide that has splintered climbers, one that is at odds with the tradition of Royal Robbins and Henry Barber. These are the sports climbers; they evolved from what the Yosemite crowd called hang-dogs, a group that had grown from the rock jocks.

In the long tradition of mountaineering and its offspring, rock climbing, one idea remained a constant: you started at the bottom and worked your way up. The idea was to challenge the unknowns of gravity, to push yourself each time you encountered a hold that might not work, to confront fairly what the rock gave you. Although standards rose each year, they weren't rising fast enough in the opinion of some who entered rock climbing in the 1970s. They decided to concentrate their effort on the crux—the most difficult part of the climb. To do it, they would walk up to the top of a cliff, anchor a rope, then let themselves down to the crux, hang there (hence, the hangdog moniker) and climb it over and over again—falling dozens of times—until they had mastered exactly the right moves to solve the problem. Once that was done, they would go to the bottom and do the entire route. This technique was then extended so that a climber might work on a hard route for months, hangdogging to the difficult sections, figuring out every hold, and then putting the whole route together in what was claimed to be the route's first free climb. For many climbers, this violated the whole idea of what a first ascent was.

Traditional climbers were outraged, yet they couldn't deny that progress came faster this way. Europe, and especially France, where sports climbing gained early acceptance, began to produce climbers whose techniques went beyond anything the Americans could do. Europeans also had no misgivings about artificial climbing walls: they became so popular that dozens were built into elementary schools, which gave aspiring climbers a chance to realize their potential at the earliest possible age. This was climbing shorn of the uncertainties of weather or rock: a pure confrontation with gravity, with all real risk gone.

The fear of falling, though, remains, but only as an instinctive reaction. "For experienced climbers," Mike told me, "the fear of falling is definitely in the back of their mind. But I wouldn't go so far as to say it's the thing that keeps them in the sport. I think the thing that keeps climbers in the sport is the movement. The difficulty of the moves, and the flow. But fear of falling definitely has an impact when you're up there climbing and you're getting pumped and there's a quick draw safety right here for you to clip but you can't even let go to clip it. If you climb past it and you don't clip it, the fall is increased every step you take. The further away you get the more aware you are of how far you could fall."[7]

For any seasoned climber, the dread of falling has been reinforced a hundred times over by close calls. I asked Mike if one stood out for him. He had no trouble remembering.

"I climbed in the Black Canyon of the Gunnison when I had climbed maybe only three months. It was my first time route finding, and our first time with bigger walls like that. We thought we were on a route that was of a given grade that we thought we could handle and we were actually two routes to the right of that on a very dangerous, very hard route. I thought I was on a 10b and I was actually on an 11dr/x."

What Mike is talking about is the way climbs are rated, and the history of the rating system shows just how far climbers have progressed beyond what was thought possible. When German mountaineer Willo Welzenbach came up with his six-level system to rate climbs in the 1920s, his scheme was the model of simplicity. The number 1 signified a walk on a trail; 2 was casual scramble where hands are needed for balance; 3 required hands at all times; 4 was steep enough so that rope protection was advised; 5 required rope protection because a fall could be fatal, and 6 meant that you could only climb by hanging from hardware pounded into the mountain.

In the early 1950s, Royal Robbins, Don Wilson, and other climbers at

Tahquitz Rock in southern California felt that the fifth level needed further articulation because there was a vast range between the easiest and the hardest of the climbs they were doing. The group divided the fifth level into nine sublevels, indicated by decimal points: 5.1 being the easiest, 5.9 being the most difficult, and 5.10 impossible, because in the decimal system, a 5.10 would equal 6—the level of artificial aid. This system, which eventually came to be called the Yosemite system, worked for only a short time—by 1952, Royal Robbins himself had climbed a 5.9. Soon others were climbing routes that were even harder. In some, a climber would support himself under an overhang by pinching the smallest nubbins in the rock with fingers that had been conditioned to become vise-like in their grip. It was climbing that was beyond 5.9 but used no hardware, and so it couldn't, in the conventional system, be rated a 6. The solution was to increase the decimal number to 5.10 and up, which violated math but accurately reflected the difficulty of the climb. Once the breakthrough was made, it was followed.

Still, as climbing progressed in smaller and smaller increments, the system needed to be tweaked once again. Climbers knew that all 5.10 climbs weren't the same, and so letter grades were appended to climbs greater than 5.10. A climb might be a 5.10a or a 5.10b, all the way up to a 5.10d. At the moment, the most difficult routes that have been successfully climbed are 50 percent harder than what climbers in the 1950s believed to be the most difficult climbs that could ever be done. Today, the top-rated climbs are rated a 5.14d.

When Mike told me he thought he was on a 10, he meant a 5.10. Mike went on to explain what a 5.11d/rx was, beginning with the "rx," which is yet another way climbs are shaded—in this case, to succinctly express what might happen if you fall. "RX is like a danger kind of grading," Mike told me. "R is kind of stolen from the movie grades. If something's kind of run out [meaning that there is a long way between points where you can anchor a rope], they give it an R rating. And if it's run out and you're going to hurt yourself or possibly kill yourself, then they give it an X rating."

I asked Mike what was going on in his mind once he had discovered he was on a much more difficult route than he expected. "I remember yelling down to my friend, 'You're sure you got me, cause I'm going to fall a long way and I don't know about this piece, if it's going to hold me.' He tried to be reassuring, 'I got ya, I got ya, okay, I got ya.' And when I fell, it was like time stood still. I knew what was happening the whole time.

"When you take a big fall like that, everything gets really quiet, and very crisp and clear in your mind. You're falling at whatever rate a human body falls, but it feels a lot slower, just because you're paying such. . . . I mean, like, fear and falling and the adrenaline rush of it all, and the . . . it's different than when you're on a bike and you're going through tricks and you're in the air and you fall, it's like instantaneous. But standing there knowing you can't hold on to the rock any more and just knowing that eventually you're going to have to let go, that aspect of fear makes it really crisp. I can picture it precisely, what it feels like. And everything kind of slows down, and then just the shock of hitting the end of the rope and slamming into the rock, it's kind of hard to describe, it's like you're safe but you're not safe. You give a quick once-over on your body to check and see if you're okay. And that particular time I realized I had sprained my ankle pretty bad. I hit the edge of the wall with my heel and rolled it. And then you kind of go into survival mode and figure out the easiest way and the best way to get down."

Paradoxically, climbers, who spend all their time trying not to fall, know more about the mental dynamics of falling than, say, skydivers, who do nothing but fall. What climbers understand is how powerful the instinct against falling is—and that it is their enemy. Physical changes—the tightness in the throat, the flush of blood, the surge of adrenaline that constricts muscles—evolved to help the body prepare for the anticipated emergency. Yet when you're hanging by fingertips, these natural reactions instantly move you closer to the catastrophe your body is trying to avoid. A climber who can't hold his physical reactions at bay will become exhausted in no time. "When you're on routes that are hard," Mike told me, "getting through and thinking you're going to fall takes more of a toll than just if you get through. If you're scared, you hold on tighter. And that's the mental game."

It is a game that is as subtle in its success as it is blunt in its failure. "It's blocking falling out of your mind and thinking more about overcoming your doubts with positive—." Here, Mike stopped himself, and the pause was telling, almost as if he were reacting to a superstition. "For me, it's not even a question of positiveness. It's more like being fully objective with myself and saying 'I will do it.' If I try to be positive, I know I'm fooling myself and trying to mask the doubt. And so I approach it objectively and say, 'Do this move,' instead of 'You can do this move.' "

I asked Mike if he had ever tried free soloing, the ropeless climbing that Henry Barber practiced, and he said that he had. "Nothing really huge," he declared, then described a rating that would have been at the pinnacle of climbing only forty years ago: "maybe like 5.11, eighty or ninety feet." Unlike Barber's belief that not knowing a route is what protects him, Mike has only free-soloed routes he knows by heart; for him, this fortifies his will against the body's alarms. "You don't think that you're going to fall, so you don't," he said, and then added quickly, "Kind of."

Trying to express the delicacy of the game, Mike next seemed to reverse himself.

"I think the fool is the person who ignores what their body is telling them to do. That's the person who ends up dying. If you're going to solo, you have to listen to your body. Listen to your heart, your mind. Your fear is the thing that keeps you alive. So it's a good thing, if you don't ignore it."

Suddenly, Mike grew serious. "I had a very good friend die soloing. Derek Hersey."

Hersey's death in 1993 stunned the climbing community. This was not a weekend climber or even a very good amateur. At thirty-six, Hersey was a sponsored professional, and one of only a handful that included Henry

Barber, Peter Croft, and John Bachar, who were considered among the best soloists in the world. Hersey's body was found at the base of Sentinel Rock, the same route Henry Barber had used in 1973 to show that big-wall soloing was possible. In an interview before his death, Hersey said of soloing, "There's nothing else that makes me feel so alive," and went on to tell why: it was the mind game, at its highest level. "When you're free-soloing, you can't afford to be distracted. You concentrate on the flow from move to move to move. You exist only in the present."[8]

Mike looked down. "The fact that he was such a master soloist and fell brings it all home to me just how incredible John Bachar and Peter Croft are. That they can last as long as they have doing what they do, where anything you do wrong results in your dying. And especially outside. It's not like somebody's up there tightening all the holds down and making sure they stay fixed to the rock. It's a natural environment. Things break, water seeps out of the rock, whatever, a bird could fly up and hit you in the chest, and you're done."

We stood for a moment looking up at the artificial wall, up to where climbing had been domesticated.

"I don't think I've soloed since then," Mike said.

I asked him why he thought soloing was so important to Derek. "It's hard to say what drives people to their parts of climbing. I prefer bouldering. All I need are my shoes and my chalk bag. Maybe that's part of what drives those guys. Very simple. Me, the rock, shoes, and a chalk bag, and that's all you need." As he kept thinking about it, he decided that his explanation was personal and that he hadn't penetrated the heart of the mystery. "I wouldn't feel comfortable even venturing a guess as to what makes them do that," he said of serious free soloists. "I'm awed by it."

Then Mike ventured a guess. "I think it's probably the rush at first, but it's also, for lack of a better term, probably a little bit of cockiness. Not outwardly, but inwardly. Knowingly putting yourself at risk just because you have the confidence that you can get out of it."

He paused, then added, "Maybe."[9]

The urge to pursue activities that most people consider highly risky, such as BASE jumping, shown here, appears to be a natural drive in a significant portion of the population. Studies have shown that these people's tastes and opinions on everything from food to politics are also far different than those who are risk-adverse. *(Copyright © Red Bull / Ulrich Grill)*

CHAPTER 12

# Sensation Seekers

It was soon apparent that sensation seeking was not just another ad hoc trait measure, but was central to a basic dimension of personality.

—Marvin Zuckerman, on his research of sensation seekers

Mike Pont's sense that free soloists have the confidence to believe they won't fall may seem obvious. It's the confidence itself that needs to be explained. Climbing a sheer cliff with no protection is so risky it seems no one in their right mind would do it, and so the explanation many people believe is that a free soloist uses bravado to mask an unconscious desire to die—in other words, Freud's idea of the death wish.

Free soloists, of course, deny this. Many claim that for them free soloing is safe for the same reason that driving a winding mountain road isn't death defying. Even though a moment of inattention could send you over the edge, basic skills reduce the risk to a safe level. "You are probably pretty confident that you could walk from here to the door without tripping," notes free-soloist Jeff Achey. "Climbing is the same for me. It's an ordinary feeling to be supported by finger tips, hanging on a rock wall. The possibility of falling feels very remote—like being struck by lightning."[1] John Bachar, one of the world's best free soloists, put it succinctly: "People free solo when they walk to the store. The real issue is the difficulty of the route."[2]

It's the matching of skill with difficulty that defines how much risk there actually is, and statistics show that free soloists as a group do a good job of making that judgement. Between 1951 and 1984, 850 people were reported to have died climbing; of these, only one was a free soloist. The reason Derek Hersey's death in 1993 was so shocking was precisely because it was so unlikely. Among the world's best free soloists, only Hersey has died since the sport's beginnings in the 1960s, even though the routes have steadily become more difficult. "When I first soloed a 5.12," Bachar has written, "it was one of the first 5.12s to be soloed in the world. Now I solo 5.12 every other day."[3]

## The Undiscovered Instinct

The idea of a death wish, at least as it relates to gravity play, is a myth, yet it continues to thrive. There seems to be something about challenging gravity, even in minor ways, that causes people to lose all perspective on what the risks actually are. A case in point is the suggestion made by the American Academy of Pediatrics that home trampolines be banned.

"About 29,600 children aged 18 or younger visited the emergency room after a trampoline-associated injury in 1990 and 58,400 did so in 1995—a 98% increase in injuries," the 1998 article in *Pediatrics*, the Academy's journal, reported. "About 93% of those accidents occurred at home. . . . Overall, 249,400 youngsters suffered trampoline-associated injuries during the six-year period, and 3% were serious enough that a child was admitted to the hospital." The author of the report, Dr. Gary Smith, concluded, "The sale of trampolines for private recreational use should be stopped and a trade-in campaign should be conducted nationally to decrease the number of existing backyard trampolines."[4]

In the article, Smith claimed that a child could bounce thirty feet (and so, fall that far), warned that studies linked trampolines with quadriplegia and death, that "spinal cord injuries with permanent paralysis were associated more frequently with trampolines than with any other gymnastic sport," and called the doubling of trampoline injuries "an epidemic."[5] Major news outlets carried the story, some enhancing their coverage with a profile of a kid who had been hurt.

Yet much of Smith's study was flawed. Home trampoline owners know that it's impossible for an adult, let alone a child, to bounce thirty feet high on a home trampoline, which doesn't have the power of a professional model. Even if it did, no one but a trained gymnast could reach that height. A six-foot bounce is tops for a skilled adult, and kids typically bounce no higher than three or four feet. As Smith admitted elsewhere in his paper, it's the serious competitors who account for the majority of catastrophic injuries and deaths; since 1990, the total number of trampoline-related deaths for children was six.

Statistically, a kid riding in a car is sixty times more likely to die than one jumping on a trampoline; a kid swimming is two hundred times more likely to die. The vast majority—97 percent—of children hurt on trampolines get scrapes, cuts, or at worst, a broken bone or dislocation. A close reading of the article reveals that the proportion of kids hospitalized from these injuries actually dropped from 5 percent in 1990 to 2 percent in 1995, which only amounts to about a thousand serious injuries a year. Trampolining's average 3 percent hospitalization rate, as it turns out, is about the same as for all pediatric injuries.

Moreover, the sole reason trampoline injuries doubled was because the number of trampolines in use doubled. When this was pointed out in a letter to the journal's editor, Smith's response was to compare trampolines to an infectious disease, as if there could be nothing good about them. And while the Academy has never advocated eliminating or even reducing the number of inline skates (76,000 emergency-room visits a year), playgrounds (250,000 a year), or bikes (more than 400,000 a year, as well as 600 deaths), it seemed comfortable suggesting that trampolines be completely eliminated.

The Academy's illogical call to ban trampolines fits the pattern of gravityphobia that has continued since Garnerin's first parachute jump.

But why does playing with gravity create such hard and fast lines of opinion? Surprisingly, science has an answer.

AFTER THE TURN of the century, when Freud's psychoanalysis and Darwin's theory of evolution showed people to be closer to animals than anyone had imagined, psychologists began to wonder exactly what instincts were and how they might work in humans. Early theories posited that an instinct could be defined as the drive an animal had to reduce the tension it felt when its basic needs weren't being met. By the 1950s, researchers were testing this idea and reached surprising results.

According to theory, a mouse who had been sated with plenty of food and sex shouldn't feel tense about anything. Therefore, it should simply sit in a box and sleep. Instead, it would go nosing around, looking for something to do. When experimenters rigged up levers that would deliver different kinds of stimulation, the mouse would go for the one that gave it the most variety, even if the stimulation had nothing to do with a known primal drive. Experiments with monkeys and people showed the same tendency. Apparently animals and humans had an instinct for something more than mere survival. They seemed to need sensation.

By the 1960s, researchers began to study this effect in humans, using social-isolation and sensory-deprivation experiments to intentionally frustrate the newly discovered sensation-seeking drive. Subjects would be locked in a room for eight hours, for example, or floated in a tank of body-temperature water.

One researcher was Marvin Zuckerman, who thought he might better understand what was going on if he could predict who would become most agitated in a sensory-deprivation experiment. His approach was based on the contemporary idea that some people need more stimulation than others just to feel normal, that every person had an optimal level of arousal (OLA) produced by an optimum level of stimulation (OLS). To try to quantify a person's OLA, Zuckerman developed a questionnaire that rated people according to what he called the Sensation Seeking Scale.

Zuckerman's questionnaire asks if you like wild parties or quiet conversation, if you like to watch movies more than once, if you would consider parachute jumping, if you like to try new foods, and thirty-six other questions along the same lines. Zuckerman found that people who scored high on his scale did, indeed, seem more uncomfortable in the tank, as shown by their "motoric restlessness"—their tendency was to move around a lot. This wasn't a surprise, but another result was.

One of the experiments, called social confinement, put two people together in a room for eight hours. While it was expected that this would drive a high-sensation seeker crazy first, exactly the opposite happened: stress levels for low-sensation seekers went up far more. This meant that sensation seeking didn't just involve physical stimulation, as was first thought. That unexpected result, and others, convinced Zuckerman that he was on to something. He quit studying sensory deprivation and began to zero in on sensation seeking. "It was soon apparent," he has written of that time, "that sensation seeking was not just another ad hoc trait measure, but was central to a basic dimension of personality and predictive of a wide variety of life experiences, behaviors, preferences, and attitudes."[6] Zuckerman was soon joined by other researchers, and the field has now grown to include hundreds of studies.

The Sensation Seeking Scale actually measures four different traits: Thrill and Adventure Seeking, Experience Seeking, Disinhibition, and Boredom Susceptibility. While the Thrill and Adventure Seeking scale has the most to do with people who bungee jump or skydive, all of the qualities tend to go together, and as far as sports go, people who fool around with gravity tend to be higher in their overall sensation-seeking score than people who practice other risky sports, such as scuba diving. In one survey, those scoring highest of all were expedition climbers, followed by parachutists, elite climbers, and whitewater canoeists.

As studies became more refined, an overall picture of the high-sensation-seeking personality began to emerge. If you enjoy flying down a mountain on a snowboard, it's likely that you also like new experience for its own sake, aren't an inhibited person, and are easily bored by repetitive tasks. It was later discovered why high-sensation seekers didn't have as tough a time locked in a room with one other person as did the lows. Because they're looking to stir things up, highs are more self-revealing and expect the same of others. That makes lows uncomfortable—it's socially dangerous. The situation is similar to one we've all seen: a reserved fellow cornered at a cocktail party by a flamboyant chatterbox who won't shut up about his latest gambling spree in Vegas.

What high-sensation seekers don't have is a death wish; on this point, Freud was conclusively proved wrong. When researchers looked at risk taking as an isolated trait, they found that just as free soloists claim, high-sensation seekers don't like risk any more than anyone else. Some psychologists guessed that this was because highs have a trait that causes them to underestimate risks—but their studies disproved that, too. Although highs drive faster, they're no less likely to buckle their seat belts; although they have sex with more partners, they're no less likely to use condoms. When researchers surveyed people who had had spinal cord injuries—including a group who was hurt during what was termed "imprudent" behavior—they thought they would find the sample filled with high-sensation seekers. They didn't. Highs are no more prone to injury than lows, and the evidence indicates that it's because highs do what any rational person would do: they reduce risk by increasing their skill.

And highs may be able to increase their skills more easily than lows. Although Zuckerman began with a simple questionnaire, his and others' research has gone on to show that a high-sensation seeker's entire nervous system, including his brain, is wired differently from that of lows. The telling evidence is from tests that directly measure the nervous system's response to different stimuli, including measurements of cortical arousal, heart rate, and skin conductivity. One casualty of the research was the Optimal Level of Arousal theory, the idea that a high-sensation seeker is in a chronic state of underarousal and so needs thrills to feel normal. Instead, electroencephalographic (EEG) measurements show that at rest, there isn't much difference between the levels of cortical activity in high- and low-sensation seekers. It's only when something happens that their reactions vary—but the variance is dramatic.

These tests give clues to a person's orienting reflex, a reaction that maximizes and clarifies the newest signals flowing in from the senses. When you're surprised by something—a loud sound was used in the tests—your brain can jump into an orienting mode or instead move directly to a defense reflex or a startle reflex; it depends on how intense the surprise is. The orienting reflex is accompanied by a drop in heart rate, while both the defense reflex and the startle reflex produce sharp jumps in heart rate. When high-sensation seekers were tested against lows using progressively louder tones, it was found that highs had a stronger orientation reflex. By the time the tone got to 80 decibels, the lows' heart rate shot up immediately, while the highs' rate went sharply down for two full seconds before rising as they entered a defense or startle reflex. (Incidentally, in tests involving anxiety, it's been shown that lows feel uncomfortable when their heart rate rises, while highs actually like it). In another test, subjects were shown a series of slides that would repeat and then suddenly change. When the change came, highs showed a big jump in their skin conductivity—another measure of the orienting reflex—compared to the lows.

What this means is that high-sensation seekers, with their stronger orienting reflexes, are better able to process information that's blasting in from their senses—say, when they're snowboarding down a mountain on the verge of losing control. A low-sensation seeker's brain skips the processing and freaks out, moving directly to a fight-or-flight response.

Because high-sensation seekers have a stronger orienting reflex, other experiments have shown that they can focus on a task more quickly and can perform it with fewer errors, even when there are distractions—and some tests show that their performance *improves* with distractions. What they aren't good at is sustained, repetitive tasks. If there's nothing new going on, their mind wanders and they start to make mistakes.

Always on the lookout for new experience, high-sensation seekers broadly scan their environment and tend to combine what they find in unexpected ways. They score slightly higher on intelligence tests and far higher on tests of cognitive innovation, variety, and originality. In general, their thinking has been found to be "divergent."

Zuckerman allows himself some speculation about why we might have evolved to have this trait. He compares it to the approach and withdrawal traits that can be observed in all species, and traces ours back to our evolution as hunters. "A successful hunter must take some risks and even enjoy predation. Because of the risks, a moderate but not too high level of the trait was probably optimal for survival, reproduction, and insuring the survival of one's offspring."[7] Although Zuckerman doesn't elaborate, it seems possible that the trait goes back even farther, to the time our ancestors spent in the trees: a successful primate had to take risks when leaping from branch to branch—and had to enjoy it—to compete against other species for food and against his own for a prospective mate.

If you like gravity thrills, some of these findings hold practical advice for your love life: you'll do better to link up with another high-sensation seeker. In one study, the subscale within the sensation-seeking scale that correlated most highly with sexual activity was the Thrill and Adventure Seeking subscale. And in another study, one of married women, Zuckerman reports, "The high sensation-seeking women preferred a higher frequency of intercourse, reported more frequent multiple orgasms, more copious vaginal lubrication during sex, more continued interest in sex during pregnancies, and even reported more sexual arousal during a laboratory session involving no explicit sexual stimulation than the low sensation seekers."[8]

If you're a high-sensation-seeking male, don't go looking for your ideal mate at the local gym. One study found that women who are rigorously dedicated to their aerobics class scored lower than average in sensation seeking.

But relationships for highs are hardly a paradise, sexual or otherwise. Low-sensation-seeking couples report higher levels of marital *and* sexual satisfaction than do high-sensation-seeking couples. And although two highs work better together than a high and a low, highs are more likely to divorce no matter who they marry.

## Thrill Gene

The essence of sensation seeking, like the talent for music or math, is mostly inherited. Studies with twins have shown that nearly 60 percent of a person's sensation-seeking traits are present at birth; environment plays a smaller role.

The primary suspect in our DNA is a stretch of genes on chromosome 11. The single gene is called D4DR, the so-called thrill gene, and consists of forty-eight base pairs. What's unusual is that we don't have just one copy of this gene, but anywhere from two to eleven copies. The number of copies we each have is a good predictor of how sensation seeking we are, although it's not absolute: some people with few copies are more sensation seeking than others with several copies. The gene seems to contribute about 10 percent to the overall heritability of sensation seeking. So far, the other genes contributing to the effect haven't been found.

D4DR is particularly interesting because it builds a receptor in the brain, called the D4 receptor, that's sensitive to dopamine. Dopamine is a chemical signal, a neurotransmitter, that's released by some neurons in the brain and absorbed by others. Dopamine signals action and elation; it's associated with the good feelings that come from sex and food. It's also released artificially by taking cocaine or amphetamines and is present at high levels in people in the midst of a manic state. A long D4DR gene chain builds *weak* dopamine receptors, so to feel its pleasant effects more dopamine is required—and there's some evidence that high-sensation seekers have greater amounts of dopamine flowing through their systems. A need for dopamine is probably one reason high-sensation seekers are more likely to take drugs.

Dopamine is only one chemical associated with high-sensation seeking. Another is serotonin, a neurotransmitter that has been linked to the trait of harm avoidance and depression. (Antidepressants such as Prozac work by inhibiting the brain's processing of serotonin, and so reduce anxiety and depression.) High-sensation seekers seem to have low levels of serotonin. There's also monoamine oxidase (MAO), an enzyme that regulates dopamine, serotonin, and another neurotransmitter, norepinephrine, by controlling their production and disposal. High-sensation seekers have low levels of MAO, which may mean they don't create these neurotransmitters as readily and don't get rid of them as fast as lows.

Rounding out the chemical stew are endorphins, which block pain (limited research indicates that high-sensation seekers have average endorphin levels); cortisol, a hormone associated with depression (high-sensation seekers are low in this); and the male hormone testosterone, which is associated with aggression and sexual behavior (highs have elevated levels; men as a group consistently score higher on the sensation-seeking scale than do women).

All this chemistry tells us something about the mechanics of how a high-sensation seeker's brain works, but it doesn't say much about what it's like to feel the urge to climb or fall.

## A Hit of Enlightenment

In his book *Descartes' Error: Emotion, Reason, and the Human Brain,* neurologist Antonio Damasio presents a model for how we feel anything and how our feelings interact with our thought process. What Damasio calls primary emotions—happiness, unhappiness, fear, anger, and disgust—can be triggered without thinking: this is the sudden fear you feel when you slip off a ladder and find you're falling. In this case, your body goes into a fear state before you're aware of it because the reaction has already been carried out in the limbic system, the most primitive part of the brain: the amygdala pushes up your heart rate, tenses your muscles, and releases neurotransmitters that prime your brain for action; the hypothalamus dumps chemicals like adrenaline into your bloodstream. Only after this happens do you consciously experience fear.

Secondary emotions are those that can be created by something you know: news that a friend has died, for example. Here, your *thoughts* of that event are what trigger the limbic system—and this is Damasio's main point. What Damasio calls Descartes' error is the idea that our brains have separate rational and irrational parts, that our modern, thinking brain and our primitive, reactive brain (and our body as well) are independent entities. They aren't. "Secondary emotions," Damasio writes, "use the machinery of primary emotions."[9] In other words, when you think of something frightening, your mind kicks your limbic system into action, just as if you had experienced the event in the physical world. And when you think of something comforting, your thoughts can help calm your primal reaction as well.

Damasio's theory offers insight into what makes falling for fun or recreation such a unique experience. In any sport, the limbic system will produce primary emotions—you might feel a burst of anxiety when a tennis ball comes flying at you, for example—but most of your emotions will come from what you're thinking. After you either hit or miss the ball, the event is temporarily over, and your thoughts have a chance to influence your limbic system.

Confrontations with gravity, though, are another story entirely. Because gravity is constant, the stimulation to the limbic system is relentless. And because our senses are so acutely attuned to every subtle shift in gravity, the stream of messages they deliver to the limbic system keep it in a steady state of arousal. That makes it hard for your thoughts to break through.

In my single bungee-jumping experience, as I waited on the platform, just looking down—stimulation from my visual sense alone—threw a switch that put my body in a state of primal fear. My thoughts were working to convince me that this was safe. Rationally, I knew it was safe; I knew it from the track record of the company. Yet none of that could fight its way through to calm my supercharged limbic system. I felt helpless, as if I was looking through a porthole into a nuclear reactor that was about to blow.

This is the game of gravity that Mike Pont and other climbers speak of: the constant tension between primitive fear and civilized thought. It is so sharp because the fear of falling was hardwired into our basic neurology as a fear that should never be ignored—while our intellect, during gravity play, vigorously attempts to do just that. The struggle between what we believe is safe and what feels risky ties thought and emotion into a knot, which surprisingly can create what some describe as a profound revelation.

In his book *Bone Games*, journalist Rob Schultheis tells of this kind of experience, what he called "a hit of enlightenment," that came on the slopes of a Colorado peak. Schultheis had little experience; he had only been rock climbing three months and made his climb without ropes or hardware. Reaching the summit hadn't been hard, but once there he found himself in trouble as a storm moved in and hail began. For reasons he did not understand—"I was not in my right mind that day; I had left logic far behind,"[10] he wrote—he decided not to descend the way he had come. "Possessed by something between panic and euphoria, dread and ecstasy, I began to traverse the ridge to the north—down the rubble of the summit, and out onto the ridgeline itself, a chockablock junk heap of granite."[11] Climbing down an overhang, he hung by his outstretched arms, reaching for a foothold—but it was slick with ice. With no way up or down, he hung until he had to let go.

He landed hard on a narrow ledge just inches from a two-hundred-foot drop and was injured: stabbed in the back by landing on his own spiked crampons, cut on his leg by his ice axe, and sliced across his shoulders by something he couldn't remember. Yet he had little choice but to get up and begin down again:

> Something happened on that descent, something I have tried to figure out ever since, so inexplicable and powerful it was. I found myself very simply doing impossible things: dozens, scores of them, as I down-climbed Neva's lethal slopes. Shattered, in shock, I climbed with the impossible sureness of a snow leopard, a mountain goat. I crossed disintegrating chutes of rock holds vanishing from under my hands and feet as I moved, a dance in which a single missed beat would have been fatal. I used bits of rime clinging to the granite as fingerholds. They rattled away into space but I was already gone, away. . . . *What I am doing is absolutely impossible,* I thought. *I can't be doing this. But I have the grace, the radiant mojo, and here I am!* . . . Looking back on it, I really cannot explain or describe properly that strange person I found inhabiting my body that afternoon. . . . The person I became on Neva was the best possible version of myself, the person I should have been throughout my life.[12]

Stories of transcendent moments are to be found everywhere in the literature of climbers; these are the gravity athletes who more than any others live on the battleground of will and instinct. And at least one psychologist has attempted to map it.

University of Chicago professor Michael Apter theorizes that we are always in one of two states, either excitement seeking or anxiety avoidance. He conceives of an entire arousal range that is experienced as either relaxation or boredom at the low end and as either excitement or anxiety at the high end. The difference between the two isn't in our bodily state; excitement and anxiety show the same signals: a quickened heartbeat, a flush of blood, a spurt of adrenaline. As has been shown in experiments, the difference depends on our interpretation of the arousal—how our higher brain functions interpret what our limbic system is telling us.

Since we can change our minds in an instant, relaxation can suddenly turn to boredom, and excitement to anxiety.

But in the case of excitement seeking, what is the trigger? Apter believes he knows. Perhaps not coincidentally, he uses the threat of falling as a metaphor, suggesting that the psychological landscape we cover when we do anything that involves risk is like a plateau near a cliff. Apter defines three regions: a safety zone, far from the edge where there is no danger of falling; a danger zone, near the edge; and a trauma zone, over the edge to injury (or in day-to-day experience, emotional trauma, such as getting fired). In the excitement-seeking state, we will try to come as near to the edge as possible because here is where the prospect of pleasant arousal is highest. Yet we can only go there so long as we feel safe, within what Apter calls a protective frame. This frame is subjective; it is like an imaginary guardrail at the edge. It is when the frame suddenly dissolves, when the guardrail gives way, that we stand naked on the edge. Our excitement-seeking state flips to become anxiety avoidance, and we run for the safety zone.

Excitement itself depends on the existence of the protective frame. "Think of a tiger in a cage," Apter writes. "Both the tiger *and* the cage are needed in order for one to experience excitement: The tiger without the cage would be frightening; the cage without the tiger would be boring. Both are necessary. In order to experience excitement, then, we need both the possibility of danger and something we believe will protect us from it."[13]

What a climber or anyone who takes on a risky gravity sport does is to intentionally play with the protective frame, like a child taunting a dog on a leash. "One buys excitement with fear," Apter says, "and the greater the cost, the better the product."[14]

Part of the reason for this is that whenever the limbic system is highly aroused, one of its jobs is to change the way the brain processes information. As neurologist Damasio writes,

The result of the neurotransmitter responses is a change in the speed at which images are formed, discarded, attended, evoked, as well as a change in the style of the reasoning operated on those images. As an example, the cognitive mode which accompanies a feeling of elation permits the rapid generation of multiple images such that the associative process is richer and associations are made to a larger variety of cues available in the images under scrutiny. The images are not attended for long. The ensuing wealth promotes ease of inference, which may become overinclusive. This cognitive mode is accompanied by an enhancement of motor efficiency and even disinhibition.[15]

So when Rob Schultheis says he saw things with extraordinary clarity and that he seemed suddenly possessed of superhuman abilities, or when Mike Pont says that when he fell, time slowed and everything became "crisp," they are not exaggerating. What they experienced is a genuinely enhanced state of consciousness that Zuckerman, Damasio, and others have identified with solid research.

Another psychologist, Mihaly Csikszentmihalyi, has taken his investigation of this state one step further by seeking ways to harness its benefits. Csikszentmihalyi coined the term "flow" to describe the state, and defines it as having eight components: the activity involved is challenging and can be accomplished through skill; it requires complete attention; it has clear goals; it provides immediate feedback; it shuts out all thoughts except those of the moment; it provides a feeling of control; it dissolves inhibiting self-consciousness; and finally, it causes time to seem to expand or contract. "In our studies," Csikszentmihalyi writes, "we found that every flow activity . . . had this in common: It provided a sense of discovery, a creative feeling of transporting the person into a new reality. It pushed the person to higher levels of performance, and led to previously undreamed-of states of consciousness."[16]

Flow can be experienced in hundreds of activities, but both Csikszentmihalyi and Apter, who subscribes to the flow model, use climbers as major examples of their principles, and this is hardly an accident. On the point of feedback, Csikszentmihalyi writes, "The climber inching up a vertical wall of rock has a very simple goal in mind: to complete the climb without falling. Every second, hour after hour, he receives information that he is meeting that basic goal."[17] In nonclimbing activities, he notes that "the feedback is often more ambiguous than the simple 'I am not falling' information processed by the climber."[18] In fact, Csikszentmihalyi uses the mental state achieved by climbers to illustrate each of his eight points of flow.

Playing with gravity may provide the most intense examples of the flow phenomenon, because gravity is the constant and unyielding force in which we evolved; it unfailingly arouses our primal brain; it lets us reliably slip in and out of our protective frame without the vagaries of a human opponent; and it unequivocally focuses all our attention on the here and now. In short, we can pull against gravity knowing the consequences and knowing it will always pull back with its timeless, unchanging power of arousal.

BY NOW, THE voluminous studies on sensation seeking reveal a yawning gap between the kind of person who always drives the speed limit and the kind who would ride an extreme roller coaster three times in a row. Zuckerman and others have cross-correlated the sensation-seeking scale with dozens of personality traits, and the results confirm much of what we intuitively know about people who like to experiment with gravity. It's not just a coincidence that Yosemite's Camp 4 climbers liked to drink and carouse, or that parachuting was developed by freewheeling show people instead of sober scientists, or that the first skateboarders who took to the air were practically delinquents. High-sensation seekers score high in tests that measure their need to be unique. They are nonconformists and put a greater value than most people on independence, inventiveness, and achievement. "Sensation seeking," Zuckerman found, "is directly related to liberal permissive attitudes and negatively related to conservative attitudes in political, religious, and sexual areas, and to authoritarianism measured as a trait."[19] Low-sensation seekers like representational art, highs like abstract; lows don't like dirty jokes, highs do (they also tend to like X-rated movies and magazines); lows go to church, highs don't; lows prefer bland music, highs like complex music such as classical, jazz, and the most extreme forms of rock, and they listen to music more often. There are more lows majoring in hard sciences and more highs in the humanities. Highs care more about aesthetics than money. They are more likely to smoke, drink, and take drugs. They even like sensation in their meals, preferring hot, sour, and crunchy food—a quality a researcher dubbed "oral-sadism," a term, no doubt, coined by a low-sensation seeker. Add to all this the fact that high-sensation seekers know how to use gravity to reach an altered state of consciousness unknown to low-sensation seekers, and you have a recipe for name calling all around.

It's no wonder that gravity play creates cultural wars.

For millennia, the concept of falling has been associated with the idea of failure, especially spiritual failure. In Western Christian tradition, Satan has been depicted as an angel who was cast out of heaven by God, and whose fall makes literal the metaphor of falling from grace. The incident of Satan's expulsion is show here in Gustave Doré's *The Fall of the Rebel Angels.*

CHAPTER 13

# The Metaphor of Falling

Our brains take their input from the rest of our bodies. What our bodies are like and how they function in the world thus structures the very concepts we can use to think. We cannot think just anything—only what our embodied brains permit. —GEORGE LAKOFF, COFOUNDER OF COGNITIVE LINGUISTICS

WITH THE SENSATION of falling resting as deeply as it does in the human psyche, you would expect our culture to be filled with falling allusions—and indeed, that is exactly the case. In my dictionary[1] there are seventy-two definitions of the word *fall*, and dozens don't refer to the act of moving up or down in space. For some you can still see the connection to gravity: there are the definitions associated with battle, such as "to fall upon; to attack," and "to fall back: to give way; recede; retreat." In both of these you can imagine soldiers literally falling on top of their victims who stumble as they flee, and the expression "to fall down on the job" has within it the same idea. The definition "to be overthrown, as a government," might have been inspired by the sight of flaming buildings collapsing to the ground. Even the meaning "to lose animation; appear disappointed, as the face: *His face fell when he heard the bad news*," could refer to a smile dropping into a scowl.

But it's curious that so many definitions can't be connected in any commonsense way to the literal action of falling. There are definitions that compare falling to losing control under the influence of an outside force, such as "to fall under a spell," "to pass into some physical, mental, or emotional condition: to fall asleep," and "to fall for: to be deceived." Here we might guess that the metaphor is saying that the outside force is as irresistible as gravity. But what do we make of definitions such as these: "to fall all over oneself, to show unusual or excessive enthusiasm," "to fall away: to withdraw support or allegiance," "to fall to pieces (emotionally)," and "to fall out; quarrel, disagree." It seems that the only universal factor here is that falling is used in a general way to express things quickly going wrong. In other languages, the story is much the same. In German, for example, the word for falling, *fallund*, shows up in *rückfallend* (backsliding), *schwerfallende* (being difficult), *verfallend* (decaying or declining), *zurückfallend* (relapsing), and *durchfallend* (failing or flunking).

The traditional explanation for such metaphorical usage is that in each case, ages ago, some clever person saw a similarity between two things or conditions and expressed it, and its originality impressed his friends, who spread it through the culture until it became a cliché or even an alternate definition, such as those above. Further, it was thought that metaphors were a useful but not really necessary side effect of language. The real basis of language was its literalness—that words represented things, states, actions, and so on.

That view of metaphor was essentially demolished in 1980 by linguistic professors George Lakoff and Mark Johnson with the publication of their seminal book, *Metaphors We Live By*.[2] "I discovered," Lakoff said in a recent interview, "that metaphor was not a minor kind of trope . . . but rather a fundamental mechanism of mind."[3]

Lakoff and Johnson abandoned contemporary linguistics, establishing a new field called cognitive linguistics, because the old theories weren't meshing with discoveries in cognitive science and neuroscience. Those fields, and others, were providing evidence that the mind processes information by using "modules," or "inference systems" that had evolved to handle specific conditions in our ancestor's formative environment. For example, all people have an innate understanding of plants. Steven Pinker, director of the Center for Cognitive Neuroscience at the Massachusetts Institute of Technology, summing up the discoveries of anthropologists Brent Berlin and Scott Atran, has written, "Universally, people group local plants and animals into kinds that correspond to the genus level in the Linnean classification system of professional biology."[4]

Lakoff and Johnson's insight was that a metaphor results when we use an intuitive concept, such as the nature of a plant, to understand something more abstract, such as a financial institution: thus, banks have "branches," can "grow," and have "roots." They argued that this process is universal and unconscious and can be traced to three fundamental types of metaphors that go from the more specific to the more general: structural metaphor (for example, change is movement), ontological metaphor (ideas are things; a state of mind is a place), and orientational metaphor (the future is ahead; the past, behind). Orientational metaphors include the idea that up is good, or more, and down is bad, or less. Cognitive linguist Zoltán Kövecses offered this list of examples in his book *Metaphor: A Practical Introduction*:

> More is up; less is down: Speak *up*, please. Keep your voice *down*, please.
>
> Healthy is up; sick is down: Lazarus *rose* from the dead. He *fell* ill.
>
> Conscious is up; unconscious is down: Wake *up*. He *sank* into a coma.
>
> Control is up; lack of control is down: I'm *on top* of the situation. He is *under* my control.
>
> Happy is up; sad is down: I'm feeling *up* today. He's really *low* these days.
>
> Virtue is up; lack of virtue is down: She's an *upstanding* citizen. That was a *low-down* thing to do.
>
> Rational is up; nonrational is down: the discussion *fell* to an emotional level. He couldn't *rise* above his emotions.[5]

This idea seems to be universal. "There are other languages [besides English] in which more is up and less is down, but none in which the reverse is true,"[6] notes Lakoff. In checking English, Hungarian, and Chinese—three unrelated languages—Kövecses found that all described happiness with "up" metaphors.[7]

In fact, the association with falling and failing is ancient. The origins of the English words *fail, fallacy, false,* and *fault* can be traced to the Latin word *fallere*, which means "to deceive" and which was probably influenced by the Greek words *sphallein*, "to cause to fall," and *phelos*, "deceitful." These Greek words, in turn, are linked to the three-thousand-year-old Sanskrit word *phálati*, which means, curiously, "it bursts." All of these seem to reflect the root "pal" and its variants, *pel, phal, sphal,* and *sphel,* which come from the mother of all Western languages, the Indo-European tongue, which is thought to have begun evolving nine thousand years ago.

Because of this fundamental metaphor, the idea of some thing or person losing value quickly was naturally compared to falling. But what are we to make of the expression *falling in love?* Cognitive linguistics has an answer. In this case, *falling* isn't used as an orientational metaphor—which would be negative—but as a structural metaphor that indicates change as motion. Laura Janda explains:

> The three types of metaphor are not entirely discrete and often collaborate in a given expression. *Falling in love*, for example, uses all three types: an orientational metaphor extending the use of *in*, an ontological metaphor identifying love as a place, and a structural metaphor that maps our understanding of physical falling onto our understanding of an initial encounter with love.[8]

Why is *falling* used instead of a more neutral term, such as *stepping*? Surely to indicate the helplessness of the person who's in love. This part of the phrase, then, borrows from the orientational metaphor's message that lack of control is down.

## Why Is Heaven Above?

In the cognitive linguistic scheme of things, God, being the greatest good, would obviously be above. But historically, of course, it wasn't explained this way. The classic rationale, which has been offered before and since, was succinctly put in the 1788 edition of the *Encyclopaedia Britannica*'s explanation for why the Greeks invested so much in Apollo:

> As Apollo is almost always confounded by the Greeks with the sun, it is no wonder that he should be dignified with so many attributes. It was natural for the most glorious object in nature, whose influence is felt by all creation, and seen by every animated part of it, to be adorned as the fountain of light, heat, and life. The power of healing diseases being chiefly given by the ancients to medicinal plants and vegetable productions, it was natural to exalt into a divinity the visible cause of their growth[9]

It is true that seeing gods in the sun, sky, moon, stars, and everything above is an extraordinarily common facet, perhaps even a universal part, of human culture. In the Western world, the Jews, Greeks, Romans, and Christians all saw their gods as ruling from a higher realm. So did the Hindus, who believed that seven heavens rose above the earth; the Japanese, whose indigenous religions include concepts of heaven; and the Chinese, whose idea of heaven dates back to at least 1500 B.C. All of these cultures also developed the idea of an underworld, some place below that was associated with evil, suffering, and often damnation. Early Buddhists were particularly enthusiastic, imagining eight major hells and sixteen minor hells, your time in any of them determined by the exact nature of what you did wrong.

With this cosmology, the idea that falling could represent a lapse in spirituality was obvious. In Islam and Zoroastrianism, the entrance to heaven is conceived of as a bridge. True believers won't have any trouble skipping across, but infidels will fall into the pits of hell. For the Greeks, the moral dimension of falling is expressed in the myth of Icarus, who along with his father Daedalus tried to fly from Crete over the sea on wings made of wax. The plan worked so well that Icarus got carried away and flew too near the sun, which melted his wings and sent him plummeting to his death.

In Christianity, the most famous fall is that of Satan, formerly Lucifer, who was cast from heaven to tumble into the pits of hell. As well known as the story is, it doesn't actually exist in the Bible. In Revelations, John describes a prophecy in which Satan, in the form of a dragon, will battle the archangel Michael, lose, and be thrown to the earth—not hell. Besides, earlier John describes the demons of the underworld as already having a "king over them, which is the angel of the bottomless pit, whose name in the Hebrew tongue is Abaddon, but in the Greek tongue hath his name Apollyon" (Rev. 9:11). Adding to the confusion, later commentators saw in Revelations not prophecy but history. To them, God had shown John the very day of Satan's earlier fall—and in that light, other biblical passages seemed to confirm the view. Luke quotes Jesus as saying he saw Satan "fall like lightning from heaven" (Luke 10:18), although Jesus doesn't describe it further. In the Old Testament, a passage from the prophet Isaiah has been interpreted as a foreshadowing of Satan's fall, even though Isaiah was clearly talking about the king of Babylon and how he would eventually be brought low by God. Because the king had boasted that he would rise as high in the heavens as the star Venus, Isaiah taunted him by saying, "How are you fallen from heaven, O Shining One, son of Dawn! How are you felled to earth, O vanquisher of nations! Once you thought in your heart, 'I will climb to the sky; higher than the stars of God I will set my throne. . . . I will match the Most High.' Instead, you are brought down to She'ol, to the bottom of the Pit" (Isa. 14:12).

When translated into Greek, "O Shining One," a reference to Venus, became "Lucifer," which some interpreters assumed was the devil's angelic name before he was cast out. *She'ol*, a vague Hebrew term meaning "grave" or "realm of the dead" was translated as "hell." Other citations seemed to confirm that Satan was once an angel, although the word *Satan* was a translation of a Hebrew word that simply meant "adversary"; thus, in many biblical passages, when God—or anybody—had an adversary, it was "Satan."

In spite of the skimpy evidence, the power of the falling metaphor took hold, and the image of someone "falling from grace" found its ultimate expression in the story of not just a mortal but an angel who became the most evil entity in the universe by way of a fall. Now the definitions of *fall* that include "to succumb to temptation," "to become unchaste," and "to lose one's innocence" make sense.

The corollary to falling being bad is the idea of rising being good. This too seems to be nearly universal. I was reminded of this when I visited the Hopi pueblo of Sichomovi in northern Arizona. The Hopi are the direct descendents of the people who walked into North America over the Bering Strait land bridge, and traditional Hopi religious beliefs have been almost entirely isolated from the Western world for perhaps thirty-five thousand years, owing in part to the Hopi's remote location.

Yet in the Hopi ceremonial reenactment of human creation, rising was equated with spiritual progress. Inside a round room called a *kiva*, there is a hole in the floor symbolizing the exit from a primitive world, and a hole in the roof, the entrance to the present world. When a stone is dropped signaling the end of the old world, the young men inside race for the ladder, climbing up the rungs as fast as possible to be the first to emerge into the new world. Moreover, Hopi Kachinas, spirit beings who visit the villages as represented by dancers in costumes, are believed to live on the San Francisco Peaks, the highest mountains visible from the Hopi mesas. Eagle feathers hold a special place in their ceremonies because these are believed to help carry prayers up to the sun god.

The importance of the sun as the inspiration for the "God and good are above" idea does make some sense when talking about agricultural peoples, such as the Hopi, whose crops depended on it. The argument is less compelling when we think of hunter-gatherers living in rainforests or on savannas. One conclusion you could draw from this view is that the conceptual metaphor of up being good didn't begin until the agricultural revolution, about ten thousand years ago. But cognitive linguistics maintains that the most elemental metaphors, such as orientational metaphors, have an elemental origin, as expressed by Lakoff:

> Spatial relations concepts in languages around the world . . . are composed of the same primitive "image-schemas," that is, schematic mental images. These, in turn, appear to arise from the structure of visual and motor systems. This forms the basis of an explanation of how we can fit language and reasoning to vision and movement.[10]

This implies that our concepts about up being good evolved along with our body, long before primitive farmers saw the sun as a benevolent god. In fact, I will argue that falling as a metaphor didn't evolve from a vertical hierarchy of value, but that the experience of falling created the hierarchy in the first place.

## Another Theory

Joseph Campbell, in exploring the origins of myth, has written,

> Certainly one force that can never have been absent from human experience . . . is gravity, which not only works continuously on every aspect of human affairs, but has fundamentally conditioned the form of the body and all its organs. The diurnal alteration of light and dark is another ineluctable factor of experience. . . . Hence a polarity of light and dark, above and below, guidance and loss of bearings, confidence and fears . . . must be reckoned as inevitable in the way of a structuring principle of human thought.[11]

These factors of experience were certainly present for all animals, but for primates, as we've already seen, gravity held a special importance. And if the clambering theory of consciousness advanced by Povinelli and Cant is right—if the first animals to become conscious were apes that began to carefully consider how to avoid falling—then gravity would have been the first thing any animal had really thought about. This idea has some interesting implications.



Picture our distant chimplike relative working its way through the canopy. Described in the most basic terms, when it clings, climbs, leaps, and balances successfully, it feels good; if it's unbalanced, misses a hand hold, or slips into a fall, it feels bad. There was nothing unusual about sudden emotions; a big cat clawing its way up your tree could bring unexpected fear, while sighting a potential mate no doubt brought a pleasurable flush. But there *was* something different about misjudging gravity, because from the primate's point of view, there was nothing external that caused it. Unlike other situations that evoked sudden terror, a slip wasn't triggered by anything "out there." The insight brought by consciousness was that the primate would have seen not only its successes and failures *but also its feelings* as a consequence of its own actions.

The way gravity works could only have reinforced the effect. Balancing its large weight on a small fulcrum like a branch, our thinking ancestor would have been subject to an instantaneous and intense feedback system: each misstep offered an unpleasant emotional jolt; every firm grip, a brief moment of elation. These rapidly changing feelings associated with being in or out of harmony with gravity may have given these creatures the first unequivocal sense of "rightness" and "wrongness" based on an individual's moment-by-moment action—in other words, the faintest *conscious* stirrings of what would become associations of good and bad, even feelings of virtue and guilt.

When our ancestors left the forests, the link between falling and peril could have become even stronger. By becoming bipedal, they gave up the more stable all-fours stance and bet everything on two feet and a good sense of balance. A primate that fell when running from a predator could face death just as a four-footed animal could; the difference was that for primates, running was far less secure, a precarious battle against the enemy gravity. Children too young to run quickly had to be carried. Imagine the concern a prehistoric mother would have about her baby tumbling from her arms as she ran from a predator. Or picture yourself as the child: you're swept up in your parent's arms, the moment supercharged with the emotion of salvation. Or you're thrust up into the branches and urged to climb as high as you could. These events must have been accompanied by a psychic imprint of enormous proportions.

Even as our ancestors evolved into meat-eating hunters and scavengers, the values associated with trees, and with higher and lower, would have been reinforced. Picture a band of *Homo erectus* moving across the savanna a million years ago. By this time they had become extremely well adapted to long overland journeys, yet there was still no reason to lose the adaptation to climbing, and much to ensure that grasping hands and loose shoulders would remain unchanged. Besides offering the ability to carry a child or throw a rock or a spear, *Homo erectus*'s tree-formed body offered one more advantage: he could always climb a tree or rock to see where he was going. Perhaps it was the duty of the leader to climb to sight herds of game or a predator's kill that might provide food. The leader would then climb down and lead the hunt. He knew the path, even if the others didn't; he could, in a way, see the future, like a god that knew more than mortals.

This habit we acquired from the trees—climbing to get a view—holds other implications about what happened when we reached the ground, as noted by ecologist and environmental philosopher Paul Shepard:

> Movement through the dense crowns of trees is not at all like moving across open spaces. There is a more emphatic streaming, a clearer before and after. Three-dimensional travel requires more decisions and perhaps better planning, in which past experience is critical. See-

> ing ahead in space was inextricably bound with seeing behind in time; seeing ahead in time was a metaphor of seeing ahead in space. By translating adjacent order to successive order, time and space became interchangeable, with an incipient cross-symbolism. When this sort of mentality was carried once again to the ground, it may have opened new vistas to thinking a widened capacity for tradition. In transmitting cultural information, "I see" means "I understand" when applied to events in the past. A "seer" is one whose vision extends forward through the unity of time and space in which the tradition-oriented society lives.[12]



As *Homo erectus* moved from the plains of Africa through Europe and Asia, climbing mountains may have provided the same sort of orienting view, and this may be one reason why mountains were, and are, held sacred in so many cultures.

## From Experience to Metaphor

Assuming that this idea has merit, when might our feelings about up and down have coalesced into abstract concepts that included rising and falling as a metaphor for achievement and failure? The beginning of language, which is the clearest demonstration of abstract thinking, has traditionally been linked with the first record of human art, such as the cave paintings and statuettes made forty thousand years ago. But many researchers today believe that the lives led by earlier hominids indicate that language probably began long before that. By 1.5 million years ago, *Homo erectus* had both good tools and the control of fire. Earlier fossils of *Homo habilis* show that in the imprints left in their skulls, there is evidence of an enlarged region of the brain that in modern humans at least, is used for language. With both these groups, it's hard to believe that animals who hunted and cooked together couldn't talk to each other.

Steven Pinker has argued for the possibility of an even earlier origin of speech, based on his research showing that the structure of language is essentially hardwired into humans, something that would be unlikely to happen over just a few thousand years. "The first traces of language could have appeared as early as *Australopithecus afarensis* (first discovered as the famous 'Lucy' fossil)," Pinker has written. "Or perhaps even earlier; there are few fossils from the time between the human-chimp split five to seven million years ago and *A. afarensis*."[13] If it is conceivable that Lucy talked, then it may have been true that she talked about falling.

By the time language had developed enough to allow myth-making—whenever that might have been along the way—there's good reason to believe that the feelings associated with falling—that down was bad and up was good—had already been in place for millions of years.

We may even have a vestigial recounting of our ancestors' reluctant exile from the safety of the forest in the story of Adam and Eve's fall. For protohumans, being far from a tree would have been a frightening thing. Climbing down to confront the dangers of predators and scarcity of food certainly resulted in a loss of innocence. It also resulted in a loss of ability. "In mythology," noted Claude Lévi-Strauss, "it is a universal characteristic of men born from the Earth that at the moment they emerge from the depth they either cannot walk or walk clumsily."[14] This may just be an allusion to the fact that newborn humans can't walk, yet it may recall earlier times, when we weren't as good on the ground as we were in the trees. Perhaps the tree of knowledge and the apple didn't originally symbolize worldly experience but ancestral wisdom and competence. Adam and Eve's banishment from the garden then—perhaps to the savanna—could be taken literally.[15]

Over the few last decades, thrill rides have taken to emphasizing the sensation of falling. Not only are rides getting higher, with the record for highest drop on a roller coaster rising from 138 feet in 1987 to 228 feet by 1991, innovations such as hanging coasters have intensified the sensation. "X," at Six Flags Magic Mountain in Valencia, California, shown here, is designed to rotate riders so they're in belly-flop position as they make the coaster's first drop of two hundred feet, falling at seventy miles per hour. *(Six Flags Magic Mountain)*

CHAPTER 14

# Falling Today and Tomorrow

There's no biodynamic test anywhere that says what a human being can stand or cannot stand. . . . If we tried to create a standard, I'm not sure it would be a thrill ride. I'm afraid the standard would be much too low.

—Harold Hudson, senior vice president of engineering and design, Six Flags Theme Parks

Today, activities that exploit the sensation of falling seem to know no bounds. Even as I write, the innovations come faster than I can research them. In the newspaper a few days ago I read that the riders of aluminum scooters are now flying off the ramps in skateboard parks; no doubt we'll see the Tony Hawk of scooters appear any day, as well as a company that builds scooters expressly for getting air. The skateboard has taken to dirt trails: there are now more than twenty companies manufacturing mountainboards—skateboards with small pneumatic tires—and an estimated half-million mountainboarders. In skydiving, the latest rage is a winged suit produced by a company called BirdMan International. It looks just as you might imagine: after exiting the plane, a diver wearing one spreads his arms and legs to reveal webbing that makes him look like a bird. Although the idea has been around since the beginning of the century, the crucial improvement that BirdMan has made is in the construction of the wings. These are inflated during a fall into the shape of an airfoil, which creates enough lift to actually carry a skydiver forward much faster than he falls. While dropping at sixty miles an hour, he can zip through the air at speeds reaching a hundred miles an hour and can actually travel up to a mile or two from where he was dropped. It's as close to flying as—well, as close as it can be until someone thinks of something else, which will probably be tomorrow.

In the meantime, established gravity sports are still pushing the boundaries. On March 7, 2000, at Lake Havasu, Arizona, Andy Macdonald attempted to break the record for the world's longest skateboard jump. He pushed off at the top of a specially prepared ramp, which dropped almost straight down, forty-seven feet into a gully that led to a short upturned ramp that would send him flying over a gap to land on another ramp. Swooping down, he then rocketed off the lip and indeed did set a world's record—fifty-six feet, ten inches.

In an ironic twist, skateboarding, surfing's kid brother, not only earned respect on its own terms but also returned to tell its older sibling that surfing could do better. Inspired by skateboarders' vert tricks, surfers learned that they too could sail six feet or so above the lip of a wave, spin in midair, and return to the water. In 1994, Ratboy, one of the new aerial breed, landed a backside 360—that is, he left the wave backward, spun all the way around, and entered backward—a trick taken directly from skateboarding. His trick was featured on the cover of *Surfer* magazine. In 1997, Christian Fletcher was crowned as the winner of the first Aerial World Surfing Championship held on Maui. "A quick glance at any surf magazine since the late '80s," writes Surfline Web site reporter Jason Borte, "proves that the aerial, and all its offshoots, are here to stay. In fact, today's sidewalk surfers are still waiting for surfing to catch up."[1] Indeed, the best vert skateboarders can now perform a 900—that's two and a half airborne spins before landing—a trick first accomplished, of course, by Tony Hawk.

But there was another frontier for surfers besides aerials: big waves. By the early 1990s, it had been believed that the largest possible waves—twenty-five feet—had been surfed. (A note about wave measurements: wave size is traditionally estimated as the size of a swell, not of the breaking wave; by the time a wave breaks, what is called a twenty-five-foot wave actually has a vertical face of about forty feet.) Waves larger than twenty-five feet couldn't be ridden, it was believed, because of three problems. First, it was extremely difficult to paddle fast enough to catch a wave that size. Second, the uprising force of the wave would most often propel a surfer back over the top. Third, if it didn't, the rider would hurtle down the face too fast to control his board—about thirty miles per hour.



Then in 1992, near Oahu's Sunset Beach, Laird Hamilton, Darrick Doerner, and Buzzy Kerbox tried a radical idea: they would eliminate paddling by being pulled with a ski rope attached to a boat speeding in front of the wave. When they were traveling fast enough to catch the wave, they would drop the rope. They soon discovered that the other two problems could be solved by redesigning their surfboards. Since they no longer needed to paddle, they didn't need a long board they could lay down on. They cut the length to under eight feet, the width to a mere sixteen inches, added foot bindings, and sharpened both ends to points. They also replaced the boat with a more maneuverable jet-ski. Records were about to be broken—decisively and fast.

On December 20, 1994, tow-in surfer David Kalama rode a wave fully ten feet higher than anything that had been ridden before. On November 23, 1995, Hamilton rode a wave with a face estimated to be an incredible sixty feet. On January 28, 1998, Ken Bradshaw rode the biggest wave yet—the face was eighty-five feet high. "Suddenly," writes Surfline's Jason Borte, "nothing was too big, and people began talking seriously about riding 100-foot waves. No longer content with baby steps, the tow-in elite has embarked on a worldwide search to see just how far we can go."[2]

The 1990s also saw an achievement in rock climbing no less impressive than surfing an eighty-five-foot wave. It happened on Yosemite's El Capitan Nose, a three-thousand-foot route so daunting that in the 1950s, Allen Steck, conqueror of Sentinel Rock, said, "We didn't think it was worth looking at. We were just intimidated by it. We never thought of even going up to test it."[3] Warren Harding did climb it in 1957. It took him forty-five days of going up and down the wall over a year and a half, drilling in hundreds of bolts and hanging ropes he would use to reach his last high point. Over the following decades, the Nose was climbed dozens of times, but always using bolts and ropes to climb over the route's harrowing sheer cliffs and overhanging roofs. Some parts of the climb had only been accomplished with a pendulum move, where a climber actually swings on a rope from a dead end across the face to a spot where the climb can be resumed.

Climbing the Nose without mechanical aid—climbing it "free," using ropes only for safety—was recognized as among the last great rock-climbing problems. By 1981, Ray Jardine, who had made the most progress, still couldn't do it, and had actually stooped to chiseling handholds into the rock, which earned him heaps of scorn from the climbing establishment.

In the early 1990s, Lynn Hill aimed at the great prize. Already a world-class climber, she had been up the route several times using aid. In 1993 she made her first free attempt with the help of her partner, Simon Nadin. She made it more than three-quarters of the way up—past the Great Roof, a massive overhang riven with only the smallest cracks that loomed over a wall that was completely blank—a pitch that had never before been free-climbed. Lynn solved the problem by hanging from her fingertips, her feet splayed straight out against the glassy face, two thousand feet above the ground. She was stopped by a section called Changing Corners. A week later, she rappelled down from the top of El Cap and spent three days working on Changing Corners until she had freed that section—another first. A few days later, Lynn and Brooke Sandahl started from the bottom and after four days of hard climbing, reached the top—Lynn having climbed the entire three thousand feet free. Astounding as that was, she returned to the Nose on September 19, 1994, for her second free attempt—this time with another record in mind. She began at night, at 10:00 P.M., and didn't stop for twenty-three hours—climbing the Nose free in a single day. Her feat has yet to be duplicated.



## The Quest for Drama

Fueled by what seems to be an unquenchable thirst for falling, it seems there will be no end to amateurs figuring out how to get something into the air that wasn't meant to be there, or to companies refining and mass-producing the objects of our ascent. But this doesn't answer an obvious question: Why now? Why, since the 1960s especially, have we become so addicted to the sensation of falling?

As extreme sports in particular have become popular, journalists have tapped the opinions of professional talking heads who've given the expected answer: that in a society made overly safe, people want to experience risk. As reported in a *Los Angeles Times* article about extreme sports, Milledge Murphy, a graduate research faculty member at the University of Florida, said this: "Historically, as any culture becomes more controlled and there are more laws, like requiring motorcycle helmets, the natural tendency is for people to express [taking risks]. Our society is preventing us from doing most things that imply risk."[4] Most other commentary runs along these same lines. The only analysis I've found to look at the question more deeply appeared in the June 1993 issue of the *Journal of Consumer Research*, and what its authors suggest adds another layer of understanding to what we know about sensation seeking and the rewards of arousal.

Richard Celsi, Randall Rose, and Thomas Leigh spent five years studying skydivers, visiting more than twenty sites in the United States and Europe and conducting a hundred interviews. Added to this was the participation of Celsi, who made more than 650 skydives himself—there would certainly be no low-sensation-seeking bias here. In essence, what the team found was that skydivers were guided by the rules of drama.

"We propose," they wrote, "that high-risk activities are dramatic in form—structured with distinct beginnings, middles, and ends—and are to a degree motivated by a dramatic worldview. Specifically, we believe that high-risk behavior is fundamentally related to our inherent dramatic enculturation. . . . Our basic proposition is that in Western society the dramatic framework is a fundamental cultural lens through which individuals frame their perceptions, seek their self-identities, and engage in vicarious or actual behaviors."[5]

But the rules of drama were laid down by Aristotle; surely something has happened lately to cause the shift. The authors point to the rise of the influence of media, and this rings true. The story of the individual quest—of the cowboy fighting single-handedly, of the detective solving the crime, of the superhero saving the world—has become the dominant schematic for Western entertainment as pumped out by Hollywood, and during the last forty years, Hollywood-style entertainment has reached out to dominate the world. The youth culture of the sixties added only one twist, and this was to laud the antihero, the one who challenges authority. Movies like *Gidget*, which directly combined antiheroes with surfing, helped spark the rise of extreme sports.

But it probably would have happened anyway, for two reasons. First, sheer demographics have compelled it. Sensation seeking peaks shortly after adolescence, and the first decade to see the beginnings of surfing, skateboarding, and rock climbing was the era of the baby boomers, a time when there were more adolescents looking for something to do than at any other time in history. They also sought sensation through protest, rock music, sex, and drugs. Their offspring, known as the baby boom echo, created the extreme sports of the 1990s, and it's likely that in an age of political cynicism, dangerous sex, and a war-on-drugs mentality, all the more of their drive began to be expressed in recreation. The second reason is the nature of gravity itself. Of all the sensations one can get through recreation, gravity offers the most bang for the buck. Gravity is free and always available; all that's usually needed is simple, cheap equipment, making it the easiest way for a sensation seeker to get a fix—and as a bonus, playing with it is guaranteed to alarm low-sensation authority figures.

But kids playing with gravity for free doesn't fill corporate coffers, and so business has had to come up with ways of delivering bigger jolts to get people to pay for it, as well as marketing hype to get people excited about paying for it, all put together in a consumable package like any modern form of entertainment.

Skydiving, for example, has been transformed from an extremely risky daredevil's stunt to something just about anyone can do on the spur of the moment and for a modest cost. Each year, Perris Valley Skydiving, which has the largest staff of skydiving instructors in the world, drops nearly eight thousand people out of planes for their first skydive. This doesn't count the thousands of jumps made by serious enthusiasts who jump on a regular basis. All in all, an accident of any kind is astronomically rare, especially among first-time jumpers. None have died during the facility's entire history of operation.

The facility, about an hour south of Los Angeles, looks like Club Med for skydivers. Along one side of what the brochure describes as a "Mediterranean Style Resort" is "The Bomb Shelter" bar and grill; next to this is "Square One," the pro shop, and farther down, a swimming pool. Men in T-shirts and women in shorts and bikini tops, many tattooed and pierced, fold their chutes on long narrow tables while others hang out, chatting and laughing.

It used to be that training was done over several days; now it takes place in about four hours and costs $299. After this, you're allowed to jump and open your own chute, although two experienced jumpers fall with you to reach for your rip cord if you don't—and there is also the added protection of a device that will automatically release the chute at a safe altitude. But if you're in a hurry, there's an easier way to go, called tandem jumping, which is also cheaper: $199. Here, after only an hour of training, you're allowed to jump out of a plane strapped to an instructor who's wearing the chute—you ride along as dead weight.

## Falling Machines

In the 1980s and 1990s, gravity thrills have also come roaring back as the top attraction at amusement parks. From the end of their golden age in the 1930s, the parks languished for decades. For most, the success of Disneyland couldn't be emulated; it was just too expensive. Then, in the 1970s, the baby boomers arrived, adolescents who cared less about landscaping and more about thrills. And just in time, several new technologies became available and would lead designers to build falling machines undreamed of before. (Ironically, Disneyland had sparked the revolution by building the Matterhorn ride in 1957, the first roller coaster to use steel tubing, although the full implications of what could be done with this new approach weren't realized then.) Just as cheap iron and Newton's physics had inspired eighteenth-century architects to build structures higher than ever, steel tubing, nylon wheels, and critically, computer-aided design inspired ride designers to create serpentine, undulating paths through space that promised a new level of falling excitement.

With the new tubular tracks, designers realized that they could completely abandon the old model that derived from a sled on a slope. They stood people up, or put them in hanging seats, letting them fling outward like so many yo-yos hung from a thrashing whip. Riding on Batman at Six Flags Magic Mountain in southern California, I was shocked at how hanging seats amplified the sensation of falling; it felt utterly unlike riding in a coaster's car. Whipping through the air, swinging from a rail overhead, my legs dangling, I didn't feel like a passenger. I felt like an ape swinging through the jungle. I am sure that this orientation, so close to what our primate ancestors had felt, acted directly on some primordial pleasure center in my brain.

Today, traditional roller coasters have given way to falling machines of tremendous variety. The renaissance has brought with it some nostalgia, and so classic wooden coasters are still being built, but the new rides are unrecognizable as coasters despite their origins. One of the most innovative is called Superman the Escape, which is also at Magic Mountain.

Superman the Escape is shaped like a gigantic letter *L*, with the sharp corner rounded a bit where it leads to the upright track. That track is four hundred feet tall. The ride is simplicity itself, at least in concept: Superman shoots riders out and up the track and then lets them free-fall back to earth.

Gravity rides like Superman are now enormous investments for the companies that operate them, and they are promoted with all the hoopla of a movie premier. Visiting on the day it was opened for a press preview, I found the patio area in front of the Superman ride rimmed with balloons and packed with several hundred journalists. Parked nearby were two TV trucks with their satellite dishes aimed skyward. A Dixieland jazz band played, and Magic Mountain staffers circulated with trays of miniquiches, minimuffins, and canapés. A forest of cameras were aimed at the podium, and many more sprouted from the shoulders of roving journalists; they were followed by sound technicians wielding padded microphone booms that looked like dustmops.

At eleven o'clock, Del Holland, president of Six Flags California, took the podium, and in a smooth southern drawl, the suited executive declared Superman a great step forward in the history of roller coasters. "And now I'd like to introduce," Del intoned as the Superman theme music began to rise all around us, "a distinguished group of first riders who have been waiting patiently aboard Superman the Escape. They are Bruce Hinds, former U.S. test pilot, and first ever to fly the B-2 Stealth Bomber; Paul Ruben, industry-known roller coaster historian and North American Editor of *Park World* magazine; and Cruz Pedregon, National Hot Rod Association funny car drag racer. And now in keeping with the Superman legend—" he paused and gestured to the ride. As the music swelled to a crescendo, the drape fell from the entrance, and fireworks—in broad daylight—burst above the ride. In the distance we now saw the car shoot straight up the vertical track, and then drop down. Then Del shouted, "Okay, now let the fun begin—let's ride!"[6]

I climbed into one of the ride's two vehicles (each holds fifteen people) ready to be enthralled by the six and a half seconds of weightlessness that was advertised to happen during the drop. To be cost-efficient, rides today have to pump hundreds of people through their turnstiles; the idea is to give them a quick, terrific jolt, and then turn them out to stagger away to the next ride—the falling equivalent of frenetic MTV editing. To that end, Superman the Escape is one of the first rides to move beyond the traditional chain lift. A chain lift would have taken far too long to crank cars up a forty-story hill, so Superman uses electromagnets attached on the car and on the track, set to repel each other. When the juice is turned on, the acceleration is instant. I heard what sounded like the rev of a huge vacuum cleaner, then bam, I was pinned to the seat like a rock in a slingshot. The building we had been in simply vanished. I could barely lift my eyelids enough to see a sign that commanded, "KEEP YOUR HEAD BACK AGAINST THE SEAT." I would have laughed if I had been able to move my mouth. The car hit a hundred miles an hour in seven seconds. Suddenly I was at the top, with only time for a glance over the side. It looked like the view from a plane. Then we fell.

And here, Superman was a disappointment, its power drained by the kryptonite of safety. Clamped into a chest-hugging harness, bolted to a seven-ton car, sliding down a rock-solid track, I felt less like I was falling and more like I was speeding backward in an overbuilt commuter train. The next thing I knew, we were gliding smoothly back into the launching dock, the magnets now putting on the brakes. The entire ride was over in thirty seconds. After that, I wandered back to the patio, hoping to talk to someone who had had a hand in designing it.

I had assumed that Superman's forces, and those of all modern rides, were precisely calculated to go just to the edge of what science knew to be the limits of what people could stand, that there was some kind of recipe to it all: put in *X* amount of G-forces (a single G equals the force of gravity) and you'll thrill 68.5 percent of the audience—something like that. In fact, it's all guesswork. I learned this talking to Harold Hudson, senior vice president of engineering and design for the Six Flags Theme Parks, and the man responsible for the team that created Superman. Harold is a cordial man in his fifties who is still a fan of good coasters. (Like me, he likes the emphasis to be on falling. He calls coasters with too many quick turns "pukers.") "There's no biodynamic test anywhere that says what a human being can stand or cannot stand," Harold told me. "The truth of the matter is, it depends very much on the human being. I think if we tried to create a standard, I'm not sure it would be a thrill ride. I'm afraid the standard would be much too low."[7] Rides like Superman are the result of people sitting around a table making it up as they go along, based somewhat on what's been done before. "You could say Superman evolved from Free Fall," Harold said.

Although it's a mere ten stories tall, Free Fall, also at Magic Mountain, is far more frightening than Superman. In Free Fall, you and one other person are instructed to sit in two chairs inside a wire cage. Then they drop the cage. No matter how prepared you are, you're not prepared. The panic is urgent, unavoidable. It's like being strapped into a plummeting elevator except that you can see through the floor. When the cage reaches the bottom, the track it's on curves to horizontal, and you end up lying on your back, with your heart pounding.

Although new rides may be based on earlier ones, designers also like to break records, and that was one of the objectives of Superman. The fastest roller coaster before Superman came close to eighty-five or ninety miles an hour on a good day (roller-coaster junkies say that hot days make the lubricants slicker). Harold and his team were out to reach a hundred miles an hour.

By its numbers, Superman should deliver the greatest falling thrills of all time. "This ride has the most changing forces of any ride," Harold beamed. "There's no ride that I know of where you get accelerated at one and a half Gs horizontally, four and a half Gs pushing you down in the seat in the curve, and then zero Gs for six and a half seconds as you go up and come back down."

As a ride of sheer thrills, Superman has its strengths. You may have felt a hundred miles an hour, but you haven't felt it in seven seconds. But as a falling ride, Superman is not a successor to Free Fall, or even a good roller coaster, and there's a simple reason why: G-forces alone don't give us the feeling of falling. Because we sense falling with the vestibular organ in our ear *and* our eyes (as well as our sense of how our limbs are arrayed in space), what we see can have a powerful effect on the overall sensation. In some circumstances, sight can even negate the signals coming from our vestibular: people in rooms filling with water on a sinking ship often don't feel the tilt of the room. Their eyes tell them that since the room is square, the water must be slanted. To really feel a fall, you need to see it. Unfortunately, Superman is built backward. You should be shot up the ramp facing backward; then when you reach the top to look down that four-hundred-foot drop, the plummet would set off the circuits in your brain like a band of screaming banshees.

WE ARE NOW in the middle of a new golden age for roller coasters—or, to be more accurate, a golden age of thrill rides. The coaster count is back up to its peak of the 1920s, with 1,332 roller coasters worldwide; of those, North America has the most, with 640.

Competition among parks is fierce. In a single year, 2001, Magic Mountain added three new coasters, which gave it bragging rights as having more roller coasters in a single park—fifteen—than anywhere else in the world. The former champion, Cedar Point in Sandusky, Ohio, countered by adding its fifteenth coaster in 2002. Parks compete not only for the quantity of rides but also for quality—which today means, more than anything else, delivering a dose of falling that dwarfs whatever the competitor's last creation could muster. Dropping two hundred feet on a thrill ride is by now practically standard, as are oblique references to extreme sports and to the X Games. In June 2002, Knott's Berry Farm in

Buena Park, California, opened the Xcelerator, a coaster whose first hill looks exactly like a vertical hairpin. Riders rise and then fall two hundred feet on inclines of ninety degrees—in other words, straight up and then straight down. Magic Mountain has launched a major campaign boasting its transformation from a theme park to "The Xtreme Park" ("Xtreme Thrills for all ages, Xtraordinary Shows, Xciting Special Events only at The Xtreme Park!" as it now crows on the home page of its Web site[8]). The park's latest ride is simply called X.

X is so extreme it almost wasn't built. At Arrow Dynamics, a Utah company that has a long record of building steel-tube coasters (including the Matterhorn at Disneyland in 1957, the first such ride), chief engineer Alan Schilke had shown his idea for something he called Fourth Dimension to colleagues, but they reacted with disbelief. Schilke's monster would put riders in seats that hung from the side of the track—and that were free to spin upside down. Not only did they think it would be impossible to build, they were certain it would make riders sick. The idea was shelved.

But when Gary Story, president of Six Flags, came calling, looking to build something to crown Magic Mountain's Xtreme Park, he was disappointed by everything Arrow showed him. He was about to walk out when Schilke decided to take a chance. He revealed to Story a rough piece of computer animation showing how Fourth Dimension would work—and Story bought it.[9]

X opened in January of 2002, and by all accounts its effect is unnerving. Unlike Superman, X floods both the visual and the physical senses. Once riders are locked into their seats, legs dangling, they leave the station backward and spend nearly a minute climbing the first long hill, watching the station shrink as a panoramic view of the park emerges—but unable to see the first drop at all. At the hill's crest, X spins riders into a belly-flop position. Suddenly, they see the ground two hundred feet below. They are dropped straight down—a ninety-degree slope—and are shortly falling at more than seventy miles per hour. Because the track is to the side, all they see below is cement. Just before being dashed to smithereens, riders are spun back to a sitting position and sail backward into a loop. Pressed into their seats, they rise inside the loop to its peak, 185 feet above the ground. Now looking forward as they speed down, they're back in the belly-flop position, just in time for a straightaway, during which they're spun into a back flip. On the next turn, the track twists, turning the train upside down like a jet fighter in a barrel roll. While upside down, riders are flipped over frontward. Then, another loop, this time facing outward. Coming out of the loop, the track again twists upside down, and the riders again flipped, this time backward, before bringing the dazed riders gliding into the station. All the loops, flips, twists, and turns after the first hill have been crammed into a harrowing forty-five seconds.

If X pushes the limits of its riders' sense of falling, it also pushes the limits of ride design, because the extreme forces X uses to slam its riders are also tearing the ride apart. Schilke's colleagues were right about one thing: X was tough to engineer. The ride's massive vehicles, twenty feet wide and seventy feet long, which hang on the side of the track, put tremendous stress on its mechanical couplings, while its complicated system for controlling the flipping of the seats (accomplished by a separate track) brings more headaches.

As Six Flags began pouring money into what would be a $15 million project, it aggressively hyped the ride's opening and then was accused of false advertising when it had to turn away furious crowds because X wasn't ready. It finally opened after a six-month delay, by which time Six Flags was withholding payments to Arrow Dynamics, as well as suing the company for $5.8 million. Even then, X was shut down several times while Six Flags crews tried to shore up the ride. Arrow couldn't help; it had filed for bankruptcy, partially because of the lawsuit and withheld payments. In June 2002, Six Flags closed X again but reopened it in August.[10] As of this writing, September 2002, X remains open, finally thrilling riders on a regular basis.

As it has for more than two hundred years, our pursuit of falling is once again leading us into the unknown. Today's extreme rides stress the human body in ways no one fully understands, and that's raising alarms. With forces that can be six and a half times the power of gravity (6.5 Gs), critics claim the new coasters press a rider's brain against the inside of his skull with enough force to cause memory loss, nerve damage, and sometimes death. "In the last 10 years," reported the *Washington Post* in May 2002, "at least 58 people have reportedly suffered brain injuries while riding roller coasters or other thrill rides at amusement parks. Eight of them died."[11]

"A roller coaster arms race has broken out in the last 10 years," stated Representative Edward Markey (D-Mass.), who has proposed federal legislation to set maximum G-force standards for rides. "The effect has been that every roller coaster today makes those of the 1950s and '60s look like a Model T."[12] Markey wants rides to generate forces no greater than 5.6 Gs for no longer than one second; if the exposure is longer, the law would require the force to be no more than 4 Gs.

Is there a whiff of gravityphobia in Markey's campaign? At this juncture, it's hard to tell. Statistically, thrill rides are among the safest forms of recreation. Joel Cliff, spokesman for the International Association of Amusement Parks and Attractions, has noted that in 2001, 320 million guests took more than 3 *billion* rides without injury, adding that "visiting a theme park today is far safer than bicycling, swimming, skiing, playing soccer, and dozens of other recreational activities."[13] Park owners maintain, with good evidence, that the majority of the miniscule number of people who are injured or die on rides have a preexisting condition that could be triggered by any number of activities. They note that signs are posted on their rides warning those with medical conditions not to board. And some of the cases listed on Markey's Web site[14] invite questions: Disney's Matterhorn ride, a very gentle coaster by almost any standard, is cited as causing a death in 2000 and a brain injury in 1964.

Adding to the confusion, often a single case is attributed to more than one ride. A 1992 fatality at Disneyland is listed as a result of "Space Mountain and Star Trek." There is no Star Trek ride at Disneyland. Assuming that the report is referring to Disney's Star Tours ride only deepens the mystery, as Star Tours isn't a coaster at all, but a motion simulator: people sit in a block of seats watching a movie, and the seats move up and down, forward and backward, no more than a dozen inches or so. In another single case, four rides are listed as culprits. If the fatal ride can't even be identified, how is it that one or any can be blamed? In his enthusiastic search for evidence, Markey seems to be grasping at straws in some cases.

Still, it is true that the parks have lobbied heavily against regulation and in 1981 succeeded in creating a loophole exempting fixed rides (as opposed to portable carnival rides) from federal oversight; that newer rides are subjecting riders to more force; and that more people have been injured in the last ten years, as thrill rides have become more extreme. Tighter regulation is probably needed—although what would be appropriate is far from clear. The question is how far should we go in limiting the enjoyment of a majority of millions to protect a minority that amounts to no more than a handful. Markey has created a panel of neurologists, amusement-industry experts, and engineers to find the answer.

IF THE HEAT is turned up on thrill rides, we may see parks building more motion simulator rides such as Star Tours, which create some of the same effects without the hazards—and this may turn out to be the future of falling machines. In generating the sensation of falling, motion simulators can be impressive.

On Star Tours, you and nine others are seated inside what looks like the cabin of a star-traveling Winnebego. Facing the front, you see a "window" to the outside, which is actually a screen. When the lights dim, a computer-controlled hydraulic mechanism synched to the action of the film starts to rumble. When the Star Cruiser begins rolling down a space station's corridor, you feel every rivet. Something goes wrong, of course, and the window shows you bursting through a door marked "Danger." On the other side is a drop down a long shaft. When the craft tips over the edge, the hydraulics boosts you up and forward perhaps a dozen inches, but combined with what you see, that's all it takes to convince you you're going down. And since the safety belts in a motion simulator can be minimal, you pitch forward and feel you're done for.

The Back to the Future ride at Universal Studios does Star Tours one better. Instead of looking out a small window, you're in a car, with windows all around, and the film is projected onto an enormous dome; it appears to be your entire reality. This makes things much worse. Naturally, there are several steep drops, and now, no matter where you look, your eyes agree with your sinking stomach.

## In the Comfort of Your Own Home

Now that computers can link film with falling, our gravity adventures are poised to reach a new level. We are at the dawn of all this, similar to when the earliest motion pictures held people in awe as they watched a locomotive simply chug down a track for a minute or two. Motion simulators showing short films are now cropping up everywhere; there's probably one in a mall not far from your home. In a few years, though, they will be *in* your home, and they will give you a lot more than a five-minute ride down a bobsled chute.

Writing in *Audio Video Interiors*, a magazine that reports on the stratosphere of home-theater installations, Keith Yates predicts that the next frontier is "haptics"—stimulating the body along with the eyes and ears. "The future is clearly the 'multi' part of multimodal," Yates writes. "The introduction of a haptic channel in a handful of cost-is-no-object residential venues in the mid-1990s marks the beginning of a transition from a bimodal experience to a trimodal one—the first time we've literally added a new dimension to entertainment since sound was added to film nearly seventy years ago. Clearly, there's a world of 'deep entertainment' around the corner."[15]

For the most part, Yates is talking about subwoofers, which have evolved from low-frequency speakers to machines that rumble so low you can feel the footfall of a celluloid dinosaur as it stomps its way toward a jeep full of terrified children. These in turn have morphed into shakers meant to vibrate a seat directly; one leading product is aptly named ButtKicker. Actual motion simulation was not part of the picture—or so I thought until I visited the 1998 Consumer Electronics Show.

In a small booth far from the giants of Sony and Toshiba, I met Michel Jacques, president of Odyssée Kinetic Home Entertainment, who invited me into his theater for a demonstration. Inside were four seats attached to a small platform. Supporting the corners of the platform were four discs, each about the size of and shape of a hockey puck, and each of these was connected to a short telescoping shaft that disappeared into a small black box. When the movie began, the platform began to roll and yaw in synch to the gyrations of the film; during a car chase down a steep hill, I was pitched forward as if in the vehicle itself.

While the Odyssée system delivers impressive results for its modest size, it isn't hard to imagine a full-blown home motion system in the near future that will heave you up and down in earnest, all synched to a movie projected onto your living room's dome-shaped surround-vision screen. When interactivity is added, along with smell, then we can escape big dinosaurs by sledding down a mountain at terrific speed, inhaling the pine forest, and kicking our legs out to stop. Or—and you can be sure we will want to—we can keep going, right off the snowy edge, sailing into the mouth of the canyon to experience the latest man-made dose of the frightening but compelling sensation of falling.

# Acknowledgments

This book began as an idea without form; it could not have been written without those who read early drafts and who helped shape that idea with invaluable suggestions: Jeff Book, Mike Hince, Scott Ray, Janice Ridenhour, Gennifer Rojas, Elizabeth Spring, Hyam Sosnow, Frank Thomas, and Diana Webster. Among those who were kind enough to spend time being interviewed are Arlo Eisenberg, Tony Hawk, Harold Hudson, Andy Macdonald, Sean Pamphilon, and Mike Pont. I am especially grateful to Jeannie Epper, Nancy Thurston, and Mat Hoffman, all of whom were more than generous with their time. I would also like to thank those who helped provide images for this book: Rochester City historian Ruth Rosenberg-Naparsteck, J. Brendan Williams at the California Historical Society, Barbara Serhant at Red Bull Photos, Ryan Snyder at SNL Communications, Andy Gallardo at Six Flags Magic Mountain, Dean Stevens at Mountain Light Photography, Heather Marsh-Rumion at Corbis, Suzanne Edmonds, and Patrick Krohn. Thanks also to Talia Botone for introducing me to Nancy Thurston.

Much of the research for this book was done using the Internet, and I would like to offer a special thanks to all who work to make the Web a rich source of information. These include institutions such as Rutgers University, which provides an excellent collection of eighteenth-century periodicals (http://harvest.rutgers.edu/projects/spectator), the Library of Congress and its American Memory Web site (http://memory.loc.gov), as well as smaller organizations such as Switchback Gravity Railroad Foundation and the Jim Thorpe Area Senior High School Computer Science Department, which together have created the Switchback Gravity Railroad Web site (www.switchbackgravityrr.org). My gratitude also extends to individuals who have created sites that offer extraordinary resources. These include Malcom Gault-Williams's astounding Web site on the history of surfing (www.legendarysurfers.com), Rictor Norton's vast resource for early-eighteenth-century newspaper reports (www.infopt.demon.co.uk), and Jim Barric's excellent site on roller-coaster designer John Miller (http://home.nyc.it.com/johnmiller/index.html), among many others. Without the work of these dedicated people, this book would have taken decades to research.

To my editor at W. W. Norton, Deirdre O'Dwyer, I am greatly indebted; her incisive suggestions made enormous improvements to the book. My sincere thanks also to Miriam Goderich at Jane Dystel Literary Management; she shepherded the book through its earliest stages and helped correct its course at many a perilous moment. And Jane Dystel, who saw promise in an unsolicited manuscript on an odd subject from a virtually unknown author but nonetheless gave it her utmost attention and support, and who also displayed extraordinary patience, I cannot possibly thank enough.

To my mother, Enid Soden, who read many early drafts, offered me an excellent perspective, and most especially, gave support when it was needed most, I will be eternally grateful.

And finally, to my wife, Kate Shein, I offer my deepest thanks. She has been all things to me: an editor who read more drafts than I could ever expect (and who is responsible for heading off some of my worst ideas); a true believer in my work who unselfishly devoted time and energy to this project at every stage; an incredibly understanding partner who endured missed social occasions and months of my inattention; and quite simply, the love of my life.

# Notes

## INTRODUCTION: BIRTH OF THE BUNGEE JUMP

David Kirke, quoted in Martin Lyster, *The Strange Adventures of the Dangerous Sports Club* (London: Do-Not Press, 1997), p. 35.

1. Irving and Electra Johnson, "South Seas' Incredible Land Divers," *National Geographic*, January 1955, p. 77.

2. Kal Muller, "Land Diving with the Pentecost Islanders," *National Geographic*, December 1970, p. 805.

3. Quotes in Lyster, *Strange Adventures of the Dangerous Sports Club*, p. 34.

4. David Kirke, quoted in Lyster, *Strange Adventures of the Dangerous Sports Club*, p. 35.

5. AJ Hackett Bungy Web site, cited August 1, 2002. Available from www.aj hackett.com/press.shtml.

6. John Miller Web site, cited August 1, 2002. Available from http://home.nyc. rr.com/johnmiller/invent.html.

7. *Rochester Observer*, November 20, 1829, quoted in Ruth Rosenberg-Naparsteck, "The Real Simon Pure Sam Patch," *Rochester History*, Vol. 52, No. 3, Summer 1991, p. 16, acquired from the Rochester History Web site, cited August 1, 2002. Available at www.rochester.lib.ny.us/~rochhist/v53_1991/v53i3.pdf.

8. *Times* of London, July 27, 1865, quoted in Fergus Fleming, *Killing Dragons: The Conquest of the Alps* (New York: Atlantic Monthly Press, 2000), p. 291.

9. Mack Sennett and Cameron Shipp, *King of Comedy* (New York: Doubleday, 1954), quoted in John Baxter, *Stunt: The Story of the Great Movie Stunt Men* (London: Macdonald, 1973), p. 23.

## CHAPTER 1: THE GRAVITY CENTURY

André Garnerin's account in *The Annual Visitor*, 1803, quoted in Peter Hearn, *The Sky People: A History of Parachuting* (Shrewsbury, England: Airlife, 1997), pp. 17–18.

1. Garnerin's assistant, quoted in Bruno Passe, "Garnerin, First Skydiver of History," Para-net.org Web site, cited March 10, 1999. Available at www.para-net.org/articles/200/indexEN.php3.

2. Leonardo da Vinci, *Codex Atlanticus*, circa 1485, quoted in Hearn, *Sky People*, p. 10.

3. Simon de la Loubè, *A New Historical Relation of the Kingom of Siam*, Vol. 2, circa 1690, quoted in Hearn, *Sky People*, p. 10.

4. This account of the Hopi ladder dance is based on Frank Waters, *The Book of the Hopi* (New York: Penguin, 1977), pp. 192–197.

5. Advertisement in the *Postman*, April 9, 1704, acquired from Rictor Norton, Early Eighteenth-Century Newspaper Reports: A Sourcebook Web site, "Contortionists and Other Performers," updated May 9, 2002; cited June 29, 2002. Available at www.infopt.demon.co.uk/grub/perform.htm.

6. Advertisement, *Daily Advertiser*, September 1754, acquired from Norton, Early Eighteenth-Century Newspaper Reports Web site.

7. *Grub-street Chronicle*, October 5, 1732, acquired from Norton, Early Eighteenth-Century Newspaper Reports Web site.

8. Johann Jakob Scheuchzer, *Itinera per Helvetiae alpinas regiones* (Leyden: Peter van der Aa, 1723, in 2 vols.), quoted in Claude Reichler, "How Dragons Disappeared from the Alps in the Mid-Eighteenth Century," abstract from *Reconceptualizing Nature, Science, and Aesthetics* (Geneva: Editions Slatkine, 1998), ed. by P. Coleman, A. Hofmann, and S. Zurbuchen. English translation by Julia Gallagher. Association for the Study of Travelers in Switzerland Web site, updated January 2002; cited August 1, 2002. Available at www.unil.ch/acvs/E/publ_0099.html.

9. Alexandre Dumas's interview with Balmat, "Jacques Balmat, dit Mont-Blanc," *Impressions de voyage en Suisse* (Paris: 1834), Vol. I, quoted in Fleming, *Killing Dragons*, pp. 44–45.

10. Ibid., p. 45.

11. Quoted in Donato Colucci, "Kemp, Pope, Falstaff and Bottom," 21st Century Shakespeare Studies Web site, posted 1997; cited August 1, 2002. Available at www.angelfire.com/ma/21stcentshakestud/kemp.html.

12. Richard Steele, *Tatler*, No. 12, 1709, acquired from the Spectator Project: A Hypermedia Research Archive of Eighteenth-Century Periodicals Web site of Rutgers University, updated June 20, 2002; cited August 1, 2002. Available at http://harvest.rutgers.edu/projects/spectator/index.html.

13. Richard Steele, *Spectator*, No. 141, August 11, 1711, acquired from the Spectator Project Web site.

14. Ralph Crotchet, letter to the editor, Richard Steele, *Spectator*, No. 258, December 26, 1711, acquired from the Spectator Project Web site.

### CHAPTER 2: TERRIFIED IMAGINATION

*Rochester Daily Advertiser*, circa 1828, quoted in Rosenberg-Naparsteck, "The Real Simon Pure Sam Patch," *Rochester History*, p. 5.

1. William L. Stone, "Visit to Niagara in 1829," Buffalo Historical Society Publications, Vol. 14, Buffalo, N.Y., 1910, quoted in Rosenberg-Naparsteck, "The Real Simon Pure Sam Patch," *Rochester History*, p. 6.

2. Ibid., p. 9.

3. The poster is reproduced in Rosenberg-Naparsteck, "The Real Simon Pure Sam Patch," *Rochester History*, p. 20.

4. Nathaniel Hawthorne, "Sketches from Memory," *New-England Magazine*, No. 9, December 1835, pp. 406–407, acquired from the American Memory Web site, updated March 1, 2000; cited August 3, 2002. Available at http://memory.loc.gov.

5. *Buffalo Republican*, October 18, 1829, quoted in Rosenberg-Naparsteck, "The Real Simon Pure Sam Patch," *Rochester History*, p. 10.

6. Flaccus, "The Great Descender," *Knickerbocker Magazine*, January and February 1840, quoted in "The Great Descender, by Flaccus," *North American Review*, Vol. 51, No. 108, July 1840, p. 231, acquired from the American Memory Web site.

7. Edgar Allan Poe, "Flaccus. — Thomas Ward," from *The Works of the Late Edgar Allan Poe*, Vol. III, 1850, p. 159, acquired from the Edgar Allan Poe Society of Baltimore Web site, cited August 3, 2002. Available at www.eapoe.org/workscriticsm/wardb.htm.

8. Ibid, p. 160.

9. Edgar Allan Poe, "Raising The Wind; Or, Diddling Considered As One Of The Exact Sciences," *Philadelphia Saturday Courier*, Vol. 13, No. 655, October 14, 1843, p. 1, acquired from the Edgar Allan Poe Society of Baltimore Web site, cited August 3, 2002. Available at www.eapoe.org/works/tales/diddlnga.htm.

10. George Stephens, *Incidents of Travel in Egypt, Arabia, Petraea, and the Holy Land* (Paris: Galigoani, August 1838), 2 vols., quoted in "Stephens's Travels in the East," *North American Review*, Vol. 48, No. 102, January 1839, p. 207, acquired from the American Memory Web site.

11. "Popular Lectures," *New Englander and Yale Review*, Vol. 8, No. 30, May 1850, p. 194, acquired from the American Memory Web site.

12. Quotes in this paragraph are from, respectively, John Hay, "Shelby Cabell," *Harper's New Monthly Magazine*, Vol. 33, No. 197, October 1866, pp. 601–602; "Last

Words of Eminent Men," selected by Sarsfieid Young, *Punchinello*, Vol. 2, No. 31, October 29, 1870, p. 75; Frederick Schwatka, "Two Expeditions to Mount St. Elias: The Expedition of 'The New York Times' (1886)," *Century; a PopularQuarterly*, Vol. 41, No. 6, April 1891, pp. 871–872; all acquired from the American Memory Web site. The book published about Sam Patch is *The Wonderful Leaps of Sam Patch* (New York: McLoughlin Brothers, 1870).

13. The broadside is reproduced in Ricky Jay, *Learned Pigs and Fireproof Women* (New York: Villard Books, 1987), p. 148.

14. *Union & Advertiser*, August 17, 1889, quoted in Rosenberg-Naparsteck, "The Real Simon Pure Sam Patch," *Rochester History*, p. 11.

15. *Rochester Gem and Ladies Amulet*, November 28, 1829, quoted in Rosenberg-Naparsteck, "The Real Simon Pure Sam Patch," *Rochester History*, p. 17.

16. *Rochester Observer*, November 20, 1829, quoted in Rosenberg-Naparsteck, "The Real Simon Pure Sam Patch," *Rochester History*, p. 16.

17. *Rochester Gem and Ladies Amulet*, January 7, 1837, quoted in Rosenberg-Naparsteck, "The Real Simon Pure Sam Patch," *Rochester History*, p. 17.

18. *Anti-Masonic Examiner*, November 17, 1829, quoted in Rosenberg-Naparsteck, "The Real Simon Pure Sam Patch," *Rochester History*, p. 14.

19. *Scientific American*, Vol. 7, No. 49, August 21, 1852, p. 386, acquired from the American Memory Web site.

20. André Garnerin's account in *The Annual Visitor*, 1803, quoted in Hearn, *Sky People*, p. 17.

21. *Sun*, September 22, 1802, quoted in Hearn, *Sky People*, p. 18.

22. The poster is reproduced in Hearn, *Sky People*, p. 16.

23. Differing versions of the conversation between Green and Cocking have been published in various sources; this one is taken from "Pleasures and Perils of Ballooning," *Harper's New Monthly Magazine*, Vol. 4, No. 19, December 1851, p. 100, acquired from the American Memory Web site.

24. *Examiner*, June 14, 1839, quoted in Hearn, *Sky People*, p. 24.

25. This according to Antony Hippisley Coxe, *A Seat at the Circus*, rev. ed. (Hamden, Conn.: Archon Books, 1980), p. 151. Coxe goes on to explain his source: "This story was told me by the great circophile, A.C. McLachlan, who had heard it from Ted and Taff Volta—themselves exceptional aerialists—who in their turn had heard it from the director of Léotard's Gymnasium in Toulouse, fifteen years after young Léotard had first thought it out."

26. Arthur J. Munby, diary, 1870, quoted in George Speaight, *A History of the Circus*, 1st ed. (San Diego: A. S. Barnes, 1980), p. 75.

27. Ibid.

28. Undated quotes from the Lockport *Chronicle* and *New York Times* in Pierre Berton, *Niagara: A History of the Falls* (New York: Kodansha America, 1997), p. 95.

29. *Punch*, November 1862, quoted in Wilton Eckley, *The American Circus* (Boston: Twayne, 1984), p. 111.

### CHAPTER 3: TROUBLE IN THE THEATERS OF GRAVITY

Undated quote from the *New York Times*, quoted in Berton, *Niagara*, p. 203.

1. Nicholas Wood, *Times* of London, quoted in Berton, *Niagara*, p. 92.

2. Nikola Tesla, *My Inventions: The Autobiography of Nikola Tesla*, ed. by Ben Johnston (Williston, VT: Hart Brothers, 1982), Chapter 6.

3. George DeWan, "A Shooting Star Comes to Long Island," Long Island: Our Story Web site, cited August 11, 2002. Available at www.lihistory.com/histpast/past0406.htm.

4. "Pedestrianism in Switzerland," *Living Age*, June 1857, p. 647, acquired from the American Memory Web site.

5. "Alpine Climbing" (from *Saint Pauls*), *Living Age*, Vol. 96, No. 1237, February 15, 1868, p. 405, acquired from the American Memory Web site.

6. Alfred Wills, *Wanderings among the High Alps* (London: Richard Bentley, 1858), p. 287, quoted in Fleming, *Killing Dragons*, p. 167.

7. Charles Hudson and Edward Shirley Kennedy, *Where There's a Will, There's a Way: An Ascent of Mont Blanc by a New Route and without Guides* (London: Longman, 1856), p. xi, quoted in Fleming, *Killing Dragons*, p. 187.

8. "Alpine Climbing," *Living Age*, p. 406.

9. "Tyndall's Glaciers of the Alps" (from the *Examiner*), *Living Age*, Vol. 67, No. 857, November 3, 1860, p. 281, acquired from the American Memory Web site.

10. Edward Whymper, *Scrambles amongst the Alps* (London: Murray, 1871), p. 125, quoted in Fleming, *Killing Dragons*, p. 227.

11. John Tyndall, *Alpine Journal*, Vol. 5, p. 331, quoted in Fleming, *Killing Dragons*, p. 233.

12. Edward Whymper, *Scrambles amongst the Alps*, excerpted in David Reuther and John Thorn, eds., *The Armchair Mountaineer* (Birmingham, Ala.: Menasha Ridge Press, 1989), p. 320.

13. Ibid, p. 325.

14. Edward Whymper, *Scrambles amongst the Alps*, p. 402.

15. *Times* of London, July 27, 1865, quoted in Fleming, *Killing Dragons*, p. 291.

16. John Ruskin, *Sesame and Lilies* (London: George Allen, 1893), p. 58, quoted in Fleming, *Killing Dragons*, p. 292.

### CHAPTER 4: SIRENS OF HEIGHT

Jenifer Warren, "Dimming the Fatal Allure of Golden Gate Bridge," *Los Angeles Times*, November 7, 1993, p. A-3. The woman Warren quoted requested that her name not be used in the article.

1. Gustave Eiffel, quoted in Henri Loyrette, *Gustave Eiffel* (New York: Rizzoli, 1985), pp. 103–178.

2. Ibid.

3. Ibid.

4. Quote from Allen Brown, *Golden Gate: Biography of a Bridge* (New York: Doubleday, 1965), p. 205.

5. Michael J. Ybarra, "Patrolling a Landmark for Troubled Souls," *Los Angeles Times*, Orange County Edition, August 20, 1996, p. E-4. Ybarra explains in the article, "Most of those who can be prevented from jumping are unlikely to try again, said Dr. Richard H. Seiden, a psychiatrist. Of 515 people who tried, but failed, to commit suicide on the bridge between 1937 and 1971, only about 10% subsequently killed themselves, according to a 1978 study published by Seiden in the journal *Suicide and Life-Threatening Behavior*."

6. "Suicide Facts," National Institute of Mental Health Web site, cited August 11, 2002. Available at www.nimh.nih.gov/research/suifact.htm.

7. "1999 Official Final Statistics U.S.A.," prepared for the American Association of Suicidology by John L. McIntosh, Ph.D., professor of psychology, Indiana University South Bend, acquired from the Indiana University South Bend Web site, cited August 11, 2002. Available at www.iusb.edu/~jmcintos/SuicideStats.html.

8. Bruce Bower, "Lethal Weapons: Gun Access and Suicide," *Science News*, Vol. 142, No. 7, August 15, 1992, p. 102(1). From the article: "From 1984 to 1987, the five counties of New York City had similar suicide rates among persons using methods equally available to all residents, such as guns, Marzuk and his co-workers report in the June *Archives of General Psychiatry.* Overall differences in suicide rates among counties, they contend, resulted mainly from varying access to lethal means other than firearms, which are subject to strict local gun-control laws. For example, fatal jumps off buildings occurred most often in the two counties with many tall residential structures."

9. Peter Fimrite, "Anti-Suicide Device for Golden Gate Officials and Passers-By Review Deterrent to Leaps," *San Francisco Chronicle*, June 10, 1998.

10. Don Sapatkin, "Why Do Some Bridges Beckon the Suicidal?" *Sunday News Journal* (Wilmington, Del.), July 8, 1984.

11. Quote from Allen Brown, *Golden Gate*, p. 200.

12. Quote from ibid., p. 205.

13. Quote from ibid., p. 218.

14. Barbara Kaufman, quoted in Michael Dougan, "Bridge Suicide Barrier Plan Has Air of Urgency," *San Francisco Examiner*, November 7, 1998.

15. "Back to the Drawing Board" (editorial), *San Francisco Chronicle*, June 11, 1998.

## CHAPTER 5: FALL OF THE NOBLE DAREDEVIL

Toni Kurz, quoted in Heinrich Harrer, *The White Spider* (Englewood Cliffs, N.J.: Prentice-Hall, 1960), excerpted in Reuther and Thorn, eds., *Armchair Mountaineer*, p. 100.

1. Annie Taylor, quoted in Berton, *Niagara*, p. 193.

2. Frank Russell, quoted in ibid., p. 197.

3. Quotes from ibid., p. 200.

4. The text of the brochure is reproduced on the A Souvineer of the Pan-American Exposition Web site, updated December 15, 1998; cited August 11, 2002. Available at http://intotem.buffnet.net/bhw/panamex/midway/midway.htm.

5. *Buffalo Evening News*, quoted in Susan Eck, "Annie Edson Taylor," Doing the Pan . . . Web site, updated August 1, 2002; cited August 11, 2002. Available at http://panam1901.bfn.org/documents/panamwomen/annietaylor.htm.

6. Frank Russell, quoted in Berton, *Niagara*, p. 203.

7. Buffalo *Courier*, quoted in ibid., p. 202.

8. Niagara *Gazette*, quoted in Berton, *Niagara*, p. 202.

9. Annie Taylor, quoted in Berton, *Niagara*, p. 206.

10. Bobby Leach, quoted in ibid., p. 205.

11. Sigmund Freud, *The Interpretation of Dreams* (1900), Chapter 6, acquired from About.com Web site, cited August 12, 2002. Available at http://psychology.about.com/library/classics/blfreud_dream6e9.htm.

12. John Tyndall, quoted in Arthur Roth, *Eiger: Wall of Death* (London: Victor Gollancz, 1982), p. 31.

13. Peter Bohen, *Grindelwald Echo*, undated, quoted in Harrer, *White Spider*, excerpted in Reuther and Thorn, eds., *Armchair Mountaineer*, p. 97.

14. *Sport* (Berne), reprinted in the *Alpine Journal*, Vol. 47, p. 379, quoted in Fleming, *Killing Dragons*, p. 350, and in Roth, *Eiger*, p. 49.

15. *Alpine Journal*, Vol. 47, p. 374, quoted in Fleming, *Killing Dragons*, p. 350.

16. Bohen, *Grindelwald Echo*, undated, quoted in Harrer, *White Spider*, excerpted in Reuther and Thorn, eds., *Armchair Mountaineer*, p. 97.

17. Quotes from the climbers in the Toni Kurz tragedy have appeared in several sources; these are taken from Roth, *Eiger*.

18. *Alpine Journal*, Vol. 5, p. 9, quoted in Fleming, *Killing Dragons*, p. 356.

19. *Living Age*, Vol. 31, No. 395, December 13, 1851, p. 502, acquired from the American Memory Web site.

20. Henry Coxwell, *My Life and Balloon Experiences*, 1887, quoted in Hearn, *Sky People*, p. 27.

21. Quotes from Hearn, *Sky People*, p. 28.

22. *Court Circular*, quoted in ibid., pp. 28–30.

23. Karl Wallenda, quoted in Eckley, *American Circus*, p. 126.

24. Dieter Schnepp, quoted in ibid., p. 133.

25. Randy McNutt, "Wallendas Fight to Survive," *Cincinnati Enquirer*, July 19, 1997.

### CHAPTER 6: RISE OF THE GRAVITY BUMS

Eddie Polo, quoted in Baxter, *Stunt*, p. 62.

1. Sennett and Shipp, *King of Comedy*, quoted in ibid., p. 23.

2. Dick Grace, *I Am Still Alive* (Chicago: Rand McNally, 1931), quoted in Baxter, *Stunt*, p. 96.

3. All quotes from Nancy Thurston and Jeannie Epper are from an interview with the author, Studio City, California, October 2, 1999.

4. Floyd Smith, quoted in Bud Selick, *Parachutes and Parachuting: A Modern Guide to the Sport* (Englewood Cliffs, N.J: Prentice-Hall, 1971), p. 7.

5. Leslie Irvin, quoted in Bud Selick, *Sky Diving: The Art and Science of Sport Parachuting*, (Englewood Cliffs, N.J.: Prentice-Hall, 1961), p. 11.

6. Randall Bose, quoted in Hearn, *Sky People*, p. 77.

7. Jack Clapp, quoted in Selick, *Sky Diving*, p. 14.

8. The brochure is reproduced on the Switchback Gravity Railroad Web site, cited July 18, 2002. Available at www.switchbackgravityrr.org.

9. O. S. Senter, quoted in "The Mt. Pisgah Inclined Plane," on the Switchback Gravity Railroad Web site. Available at www.switchbackgravityrr.org/sbrideps.htm.

10. Ibid.

11. T. L. Mumford, 1887, quoted in "The Mt. Jefferson Crossover," on the Switchback Gravity Railroad Web site. Available at www.switchbackgravityrr.org/sbrideou.htm.

12. John Miller, *Catalog*, 1923, quoted in Jim Barrick, "John Miller's Inventions," John A. Miller, Roller Coaster Designer and Builder Web site, cited August 13, 2002. Available at http://home.nyc.rr.com/johnmiller/invent.html.

13. Ibid.

14. Spokesman, quoted in Walter Lerch, "Park Ride Closed after 3 Are Hurt," *Cleveland Plain Dealer*, July 26, 1946, acquired from John A. Miller, Roller Coaster Designer and Builder Web site. Available at http://home.nyc.rr.com/johnmiller/danger.html.

### CHAPTER 7: YOUNG GODS BRONZED WITH SUNBURN

Mike Doyle, "Morning Glass: The Adventures of Legendary Waterman Mike Doyle" (Three Rivers, Calif.: Manzanita Press, 1993), quoted in Matt Warshaw, "Surfriders: In Search of the Perfect Wave" (New York: Collins Publishers, 1997), p. 123.

1. James King, *Journal*, 1779, quoted in Malcom Gault-Williams, "Surfing's Darkest Days," Vol. 1, Chapter 3, Version 2, Legendary Surfers Web site, updated August 7, 2002; cited August 13, 2002. Available at www.legendarysurfers.com.

2. Malcom Gault-Williams, "Ancient Hawaiian Surfboards and Culture," Vol. 1, Chapter 2, Version 1, Legendary Surfers Web site.

3. W. R. S. Ruschenberger, *Narrative of a Voyage Around the World*, 1838, quoted in Gault-Williams, "Surfing's Darkest Days," Legendary Surfers Web site.

4. Hiram Bingham, 1847, quoted in Gault-Williams, "Surfing's Darkest Days," Legendary Surfers Web site.

5. Nathaniel Emerson, 1892, quoted in ibid.

6. Jack London, *The Cruise of the Snark*, 1911, quoted in Malcom Gault-Williams, "The Revival, 1903–1915," Vol. 1, Chapter 4, Version 2, Legendary Surfers Web site.

7. Amateur Athletic Union, quoted in Malcom Gault-Williams, "Duke Pao Kahanamoku" (PDF Web document), p. 9, Legendary Surfers Web site.

8. Duke Kahanamoku, quoted in Malcom Gault-Williams, "The Early Redwood Years," Vol. 1, Chapter 6, Version 2, Legendary Surfers Web site.

9. Duke Kahanamoku, quoted in Gault-Williams, "Duke Pao Kahanamoku" (PDF Web document), p. 19, Legendary Surfers Web site.

10. Sam Reid, quoted in Gault-Williams, "The Early Redwood Years," Legendary Surfers Web site.

11. Tom Blake, quoted in Malcom Gault-Williams, "Tom Blake (1902–1994)," Vol. 1, Chapter 7, Legendary Surfers Web site.

12. Ibid.

13. Tom Reid, quoted in ibid.

14. *The Honolulu Advertiser*, March 10, 1926, p. 1, quoted in Gault-Williams, "The Early Redwood Years," Legendary Surfers Web site.

15. Carin Crawford, "Waves of Transformation," La Jolla Surfing Web site, updated October 16, 2000; cited August 13, 2002. Available at http://facs.scripps.edu/surf/wavesof.html.

16. The event of Gidget's naming has been described in a number of books and publications. Mine is taken from Bruce Savage, "Malibu's Uptown Surf Club '56," *H20* magazine, Spring 1980, pp. 13–14, acquired from Terry "Tubesteak" Tracy Web site, cited August 13, 2000. Available at www.tubesteak.org/magazinesa1a.html.

17. The event is recalled by Bruce "Snake" Gabrielson in "The Evolution of the Short Board," "Snake" Gabrielson Surf Page Web site, cited August 13, 2002. Available at www.blackmagic.com/ses/surf/evolution.html.

18. Tony Alva interview, *Skateboarder*, February 1977.

19. Tony Alva interview, *Skateboarder*, July 1978.

## CHAPTER 8: WALL RATS

Warren Harding, reported in *San Francisco Chronicle*, November 19, 1970, quoted in Steve Roper, *Camp 4: Recollections of a Yosemite Rockclimber* (Seattle: Mountaineers, 1994), p. 228.

1. Warren Harding, "Reflections of a Broken-Down Climber," in *Ascent 1971*, ed. by Steve Roper and Allen Steck (San Francisco: Sierra Club Books, 1971), excerpted in Reuther and Thorn, eds., *Armchair Mountaineer*, p. 86.

2. Ibid.

3. Warren Harding, *Downward Bound—A Mad Guide to Rock Climbing* (Birmingham, Ala.: Menasha Ridge Press, 1975), quoted in Gary Arce, *Defying Gravity: High Adventure on Yosemite's Walls* (Berkeley, Calif.: Wilderness Press, 1996), p. 86.

4. Ibid., p. 87.

5. Ibid.

6. Harding, "Reflections of a Broken-Down Climber," excerpted in Reuther and Thorn, eds., *Armchair Mountaineer*, p. 88.

7. T. M. Herbert, "Comment on Two Ascents of the Wall of the Morning Light," American Alpine Journal, 1971, p. 361, quoted in Arce, *Defying Gravity*, p. 89.

8. John Long, *Rock Jocks, Wall Rats, and Hang Dogs: Rock Climbing on the Edge of Reality* (New York: Fireside, 1994), p. 120.

9. Ken Wilson, editorial, *Mountain*, May 1971, quoted in Roper, *Camp 4*, p. 230.

10. Harding, *Downward Bound*, quoted in Roper, *Camp 4*, p. 230.

11. Royal Robbins, quoted in Arce, *Defying Gravity*, p. 89.

12. Harding, quoted in ibid., p. 91.

13. Henry Barber, quoted in Chip Lee, with David Roberts and Kenneth Andrasko, *On Edge: The Life and Climbs of Henry Barber* (Boston: Appalachian Mountain Club, 1982), p. 21.

14. Long, *Rock Jocks, Wall Rats, and Hang Dogs*, p. 113.

15. Barber, quoted in Lee, *On Edge*, p. 175.

16. Ibid., p. 181.

17. Ibid., p. 174.

18. Ibid., p. 185.

## CHAPTER 9: THE ARBOREALISTS

David Lambert, *The Field Guide to Early Man* (New York: Facts on File, 1987).

1. E. J. Gibson and R. D. Walk, "Visual Cliff," *Scientific American*, Issue 202, No. 1960, pp. 64–71.

2. Robert W. Grossman, "A Role the Basic Orientation System May Play in Infancy," in *Perception in Everyday Life*, ed. by S. Howard Bartley (New York: Harper and Row, 1972), pp. 33–34.

3. Povinelli and Cant's theory is detailed in Karen Wright, "The Tarzan Syndrome," *Discover*, November 1996, pp. 88–98.

4. Fred Spoor, Bernard Wood, and Frans Zonneveld, "Implications of Early Hominid Labyrinthine Morphology for Evolution of Human Bipedal Locomotion," *Nature*, Vol. 369, June 23, 1994, pp. 645–648.

## CHAPTER 10: GOT AIR?

"Bicycle Stunt History," X Games press materials, ESPN, New York, June 1997.

1. Rich Feinberg, quoted in "X Games History," X Games press materials, ESPN, New York, June 1997.

2. Jeff Ruhe, presentation at X Games press event, San Diego, June 1998.

3. Quote from "Bicycle Stunt History," X Games press materials, ESPN, New York, June 1997.

4. Mat Hoffman, television interview, X Games, ESPN, New York, broadcast June 1997.

5. Ibid.

6. Mat Hoffman, interview with author, San Diego, June 17, 1998.

7. Lane Bowers, interview with author, San Diego, June 17, 1998.

## CHAPTER 11: FALLING FOR ALL

Arlo Eisenberg, television interview, X Games, ESPN, New York, broadcast June 1997.

1. Arlo Eisenberg, television interview, X Games, ESPN, New York, broadcast June 1997.

2. Sean Pamphilon, interview with author, San Diego, June 17, 1998.

3. Tony Hawk, interview with author, San Diego, June 17, 1998.

4. Chris Fowler, X Games broadcast, ESPN, June 1998.

5. Sean Pamphilon, X Games broadcast, ESPN, June 1998.

6. Andy Macdonald, interview with author, San Diego, June 17, 1998.

7. Mike Pont, interview with author, San Diego, June 17, 1998.

8. Derek Hersey, quoted in Clifford May, "The Town That Can't Sit Still," *New York Times Magazine*, November 3, 1991, p. 56.

9. Mike Pont, interview with author, San Diego, June 17, 1998.

## CHAPTER 12: SENSATION SEEKERS

Marvin Zuckerman, *Behavioral Expressions and Biosocial Basis of Sensation Seeking* (New York: Cambridge University Press, 1994).

1. Jeff Achey, interview, *Newsweek*, October 15, 1984, quoted in Michael J. Apter, *The Dangerous Edge: The Psychology of Excitement*, (New York: Free Press, 1992), p. 35.

2. John Bachar and Steve Boga, *Free Climbing with John Bachar* (Mechanicsberg, Penn.: Stackpole Books, 1996), p. 48.

3. Ibid., p. 50.

4. Gary A. Smith, "Injuries to Children in the United States Related to Trampolines, 1990–1995: A National Epidemic," *Pediatrics*, Vol. 101, No. 3, March 1998, pp. 406–412.

5. Ibid.

6. Marvin Zuckerman, *Behavioral Expressions and Biosocial Basis of Sensation Seeking* (New York: Cambridge University Press, 1994), p. 373.

7. Ibid., p. 287.

8. Ibid., p. 192.

9. Antonio R. Damasio, *Descartes' Error: Emotion, Reason, and the Human Brain* (New York: G. P. Putnam's Sons, 1994).

10. Rob Schultheis, *Bone Games: Extreme Sports, Shamanism, Zen, and the Search for Transcendence* (New York: Breakaway Books, 1996), p. 7.

11. Ibid.

12. Ibid., pp. 10–12.

13. Apter, *Dangerous Edge*, p. 27.

14. Ibid., p. 39.

15. Damasio, *Descartes' Error*, pp. 163–164.

16. Mihaly Csikszentmihalyi, *Flow: The Psychology of Optimal Experience* (New York: HarperPerennial, 1991), p. 74.

17. Ibid., p. 54.

18. Ibid., p. 55.

19. Marvin Zuckerman, *Behavioral Expressions and Biosocial Basis of Sensation Seeking* (New York: Cambridge University Press, 1994), p. 373.

### CHAPTER 13: THE METAPHOR OF FALLING

George Lakoff, interview with John Brockman, "Philosophy in the Flesh: A Talk with George Lakoff," Edge Web site, cited August 14, 2002. Available at www.edge.org/3rd_culture/lakoff/lakoff_p1.html.

1. *Random House Dictionary of the English Language*, 2d ed., unabridged (New York: Random House, 1987).

2. George Lakoff and Mark Johnson, *Metaphors We Live By* (Chicago: University of Chicago Press, 1980).

3. Lakoff, "Philosophy in the Flesh," Edge Web site.

4. Steven Pinker, *The Language Instinct: How the Mind Creates Language* (New York: HarperCollins, 1994), p. 440.

5. Zoltán Kövecses, *Metaphor: A Practical Introduction* (New York: Oxford University Press, 2002), p. 36.

6. George Lakoff, "The Contemporary Theory of Metaphor," *Metaphor and Thought*, 2d ed., ed. by Andrew Ortony (New York: Cambridge University Press, 1993).

7. Kövecses, *Metaphor*, pp. 163–165.

8. Laura Janda, "Cognitive Linguistics," Indiana University Web site, updated August 13, 2002; cited August 14, 2002. Available at www.indiana.edu/~slavconf/SLING2K/pospapers/janda.pdf, p. 12 (footnote).

9. "Apollo," *Encyclopaedia Britannica*, 3rd ed., 1788–1797, excerpted in *The Treasury of the Encyclopaedia Britannica*, ed. by Clifton Fadiman (New York: Viking, 1992), p. 260.

10. Lakoff, "Philosophy in the Flesh," Edge Web site.

11. Joseph Campbell, *The Masks of God: Primitive Mythology* (London: Secker & Warburg, 1960), p. 57.

12. Paul Shepard, *The Only World We've Got: A Paul Shepard Reader* (San Francisco: Sierra Club Books, 1996), p. 15.

13. Pinker, *Language Instinct*, p. 363.

14. Claude Lévi-Strauss, *Structural Anthropology* (New York: Basic Books, 1963), p. 215.

15. In an endnote in his book *The Hand: How Its Use Shapes the Brain, Language, and Human Culture* (New York: Vintage, 1999), p. 361, Frank R. Wilson mentions a similar idea: "Misia Landau, in *Narratives of Human Evolution* (New Haven: Yale University Press, 1991), with or without intending to do so, injects a new candidate for the moral of the Book of Genesis by offering another justification for the expulsion of Adam and Eve fron the Garden of Eden. Another Eden existed—not the terrestrial habitat of Judaeo-Christian theology, but an arboreal one. In Landau's version of the Fall, we find another origin for the timeless theme of the vulnerability of human fortunes: tree-dwelling pre-biped human ancestors, not humans, were expelled from the garden; the reason for expulsion was neither sin nor the forbidden thirst for divine knowledge—the habitat had changed and survival in the trees was no longer an option." In my reading of Landau, her point is that the story of the Fall was a metaphor unconsciously adopted by some anthropologists that influenced their placement of bipedalism in the timeline of evolution. Wilson seems to be elaborating on this idea, suggesting that the Fall was perhaps reconceptualized by these anthropologists as similar to humans leaving the trees. My idea is that the story of the Fall is an actual ancient memory of this event—an idea, of course, that is pure conjecture and could never be proved. Still, it is interesting that the origin story recorded in the Bible is called a "fall,"

has a tree representing both paradise and the source of knowledge, an exodus from an arboreal area, and as a consequence, a new exposure to peril and pain.

## CHAPTER 14: FALLING TODAY AND TOMORROW

Harold Hudson, interview with author, at press event at Six Flags Magic Mountain in Valencia, California, March 14, 1997.

1. Jason Borte, "Skateboarding," Surfline Web site, cited August 14, 2002. Available at http://content.surfline.com/sw/content/surfaz/skateboarding.jsp.

2. Jason Borte, "Tow Surfing," Surfline Web site. Available at http://content.surfline.com/sw/content/surfaz/tow_surfing.jsp.

3. Allen Steck, quoted in Arce, *Defying Gravity*, p. 43.

4. Milledge Murphy, quoted in Jeannine Stein, "Extremists Extraordinaire," *Los Angeles Times*, July 24, 1995, Home Edition Section, p. E-1.

5. Richard L. Celsi, Randall L. Rose, and Thomas W. Leigh, "An Exploration of High-Risk Leisure Consumption through Skydiving," *Journal of Consumer Research*, Vol. 20, No. 1, June 1993.

6. Del Holland, press event, Valencia, Calif., March 14, 1997.

7. Harold Hudson, interview with author, Valencia, Calif., March 14, 1997.

8. Six Flags Magic Mountain Web site, cited August 4, 2002. Available at http://www.sixflags.com/parks/magicmountain/home.asp.

9. Reported on the Ultimate Roller Coaster Web site, cited August 13, 2002. Available at www.ultimaterollercoaster.com/coasters/reviews/x.

10. Reported on Coaster Force Web site, cited August 13, 2002. Available at www.coasterforce.com/jun2002.htm.

11. Suz Redfearn, "The Thrill Is . . . Deadly?" *Washington Post*, May 21, 2002.

12. Edward Markey, quoted in ibid.

13. Joel Cliff, quoted in Kathleen Johnston Jarboe, "Roller Coaster Deaths Spark Interest in Federal Oversight of Theme Parks," *Capital News Service*, May 3, 2002. Available at www.macon.com/mld/macon/news/nation/3189898.htm.

14. Rep. Edward Markey Web site, cited August 13, 2002. Available at www.house.gov/markey/iss_parkrides.htm.

15. Keith Yates, "Hooked on Haptics," *Audio Video Interiors*, October 1996, p. 30.

# Select Bibliography

## BOOKS

Apter, Michael J. *The Dangerous Edge: The Psychology of Excitement.* New York: Free Press, 1992.

Arce, Gary. *Defying Gravity: High Adventure on Yosemite's Walls.* Berkeley, Calif.: Wilderness Press, 1996.

Baxter, John. *Stunt: The Story of the Great Movie Stunt Men.* London: Macdonald, 1973.

Benson, Joe. *Souvenirs from High Places.* Seattle: Mountaineers, 1998.

Berton, Pierre. *Niagara: A History of the Falls.* New York: Kodansha America, 1997.

Brooke, Michael. *The Concrete Wave: The History of Skateboarding.* Toronto: Warwick, 1999.

Csikszenthmihalyi, Mihaly. *Flow: The Psychology of Optimal Experience.* New York: HarperPerennial, 1991.

Damasio, Antonio R. *Descartes' Error: Emotion, Reason, and the Human Brain.* New York: G. P. Putnam's Sons, 1994.

Eckley, Wilton. *The American Circus.* Boston: Twayne, 1984.

Fleming, Fergus. *Killing Dragons: The Conquest of the Alps.* New York: Atlantic Monthly Press, 2000.

Gregory, R. L. *Eye and Brain: The Psychology of Seeing,* 2d ed. New York: World University Library, 1966.

Hearn, Peter. *The Sky People: A History of Parachuting.* Shrewsbury, England: Airlife, 1997.

Hill, Lynn, with Greg Child. *Climbing Free: My Life in the Vertical World.* New York: W. W. Norton, 2002.

Jones, Chris. *Climbing in North America.* Seattle: Mountaineers, 1997.

Kövecses, Zoltán. *Metaphor: A Practical Introduction.* New York: Oxford University Press, 2002.

Lee, Chip, with David Roberts and Kenneth Andrasko. *On Edge: The Life and Climbs of Henry Barber.* Boston: Appalachian Mountain Club, 1982.

Long, John. *Rock Jocks, Wall Rats, and Hang Dogs: Rock Climbing on the Edge of Reality.* New York: Fireside, 1994.

Lyster, Martin. *The Strange Adventures of the Dangerous Sports Club.* London: Do-Not Press, 1997.

Mitchell, Richard G., Jr. *Mountain Experience: The Psychology and Sociology of Adventure.* Chicago: University of Chicago Press, 1983.

Pinker, Steven. *The Language Instinct: How the Mind Creates Language.* New York: HarperCollins, 1994.

Reuther, David, and John Thorn, eds. *The Armchair Mountaineer.* Birmingham, Ala.: Menasha Ridge Press, 1989.

Roper, Steve. *Camp 4: Recollections of a Yosemite Rockclimber.* Seattle: Mountaineers, 1994.

Roth, Arthur. *Eiger: Wall of Death.* London: Victor Gollancz, 1982.

Schmidt, Robert F., ed. *Fundamentals of Sensory Physiology.* Heidelberg: Springer-Verlag, 1978.

Schultheis, Rob. *Bone Games: Extreme Sports, Shamanism, Zen, and the Search for Transcendence.* New York: Breakaway Books, 1996.

Selick, Bud. *Parachutes and Parachuting: A Modern Guide to the Sport.* Englewood Cliffs, N.J.: Prentice-Hall, 1971.

Selick, Bud. *Sky Diving: The Art and Science of Sport Parachuting.* Englewood Cliffs, N.J.: Prentice-Hall, 1961.

Stanley, Steven M. *Children of the Ice Age: How a Global Catastrophe Allowed Humans to Evolve.* New York: Harmony Books, 1996.

Vermuelen, James P., ed. *Mountain Journeys: Stories of Climbers and Their Climbs.* Woodstock, N.Y.: Overlook Press, 1989.

Warshaw, Matt. *Surfriders: In Search of the Perfect Wave.* Del Mar, Calif.: Tehabi Books, 1997.

Zuckerman, Marvin. *Behavioral Expressions and Biosocial Basis of Sensation Seeking.* New York: Cambridge University Press, 1994.

## ARTICLES

Johnson, Irving and Electra. "South Seas' Incredible Land Divers." *National Geographic*, January 1955.

Muller, Kal. "Land Diving with the Pentecost Islanders." *National Geographic*, December 1970.

## ELECTRONIC SOURCES

Rictor Norton, *Early Eighteenth-Century Newspaper Reports: A Sourcebook*, www.infopt.demon.co.uk/grub/perform.htm.

Malcom Gault-Williams, Legendary Surfers, www.legendarysurfers.com.

Rochester History, www.rochester.lib.ny.us.

Spectator Project: A Hypermedia Research Archive of Eighteenth-Century Periodicals Web site of Rutgers University, http://harvest.rutgers.edu/projects/spectator/index.html.

American Memory, http://memory.loc.gov.

# Index

Page numbers in *italics* refer to illustrations.